GEOLOGY OF ARIZONA

Dale Nations
Geology Department
Northern Arizona University
Flagstaff, Arizona

Edmund Stump
Geology Department
Arizona State University
Tempe, Arizona

KENDALL/HUNT
PUBLISHING COMPANY
Dubuque, Iowa

Cover painting Cañon Cristalino #9, copyright 1981 by Cynthia Bennett. Original is acrylic on canvas, 152 × 101 cm.

Printed in the United States of America
10 9 8 7 6 5 4

CONTENTS

FOREWORD

Someone once wrote that Arizona is geology by day and astronomy by night. At every turn there is another outcrop where the stoney skeleton of an ancient era is eroded clean and bare, waiting to reveal its history. Back in the Midwest, you can look to the farthest horizon without seeing a rock, a hill, or any other clue to the history of the land. In Arizona there is always a hill or canyon on the horizon, and each city and town has its own mountain marking it as a unique and specific place, whether Camelback Mountain in Phoenix, the Catalinas at Tucson or the great volcano above Flagstaff. If you look beneath the whitewashed school monogram on the hillside, there is much to be learned.

Geology of Arizona is both a textbook and a guidebook. In a succinct and readable way it takes us through the basics of minerals, stratigraphy, paleontology and tectonics. The singular contribution of this book, however, is the way in which it teaches universal principles by use of concrete Arizona descriptions, photographs and illustrations. Arizona's turbulent geological past pours out across these pages. There seems to be no time period, no rock type, no structure (even the classic American Meteor impact site) that can't be found somewhere in the state. The authors have located them all and explained them well. This book ought to be in the library, or better, in the glove compartment, of every Arizonan.

Bruce Babbitt
Governor of Arizona

PREFACE

Humans are inquisitive creatures, constantly wondering about their surroundings, the course of human events, their relationship to other people, governments, and about life itself. Most of our daily experiences are ephemeral ones and we realize that life consists of our constantly adjusting to the needs, desires and demands placed upon us by ourselves, our families, friends, and a dynamic social order. To many of us, the opposite to our hectic daily routine is epitomized by the land we live on, a solid, steady, unchanging standard of reference. For centuries, people have retreated to the countryside, the mountains, the seashore, even the city parks, to enjoy a respite from daily activities in the solitude of natural things.

But even in the natural world human curiosity prevails and we begin to wonder about the land. What has caused the differences in landscapes, the different kinds and colors of rocks, the deep canyons and steep cliffs? Why do we find fossils of marine animals in rocks at the North Rim of the Grand Canyon, 2400 meters (8000 feet) above sea level? Why is Arizona the main copper-producing state? Why do we have so little oil production? This book is written to provide answers to such questions and to enhance the appreciation and enjoyment of Arizona for those who visit or live here.

Arizona is well known for its scenic beauty. The Grand Canyon, Monument Valley, Canyon DeChelly, the Rim country, the southern deserts, all attract tourists from other states and other countries. That scenic beauty is due, of course, to the colorful rocks and interesting land forms.

But the rocks offer more than aesthetic pleasure; they also contain clues to the history of this small part of the earth that we call Arizona. They contain important raw materials—copper, gold, silver, asbestos, even oil and gas which are of increasing importance in the modern world. They also contain certain minerals that tell us how old the rocks are, and other minerals that tell us how deeply buried the rocks were when they were formed.

Evidence of earth movements is common in Arizona. Earthquakes are relatively frequent, and numerous faults, especially in the southern and western parts of the state, attest to the dynamic evolution of the crust in past times. Geologists are still working to unravel the complex history of Arizona's rocks and landscapes. There is much debate about how and when the Colorado Plateau of northern Arizona became so distinctly separate from southern and western Arizona, why it is so much higher, and why the rocks are relatively undeformed.

We will discuss the geology of Arizona beginning with some basic concepts and principles that are necessary to establish the technique of interpreting the history of any part of the earth's crust. The book is organized with an assumption that the reader has no formal training in geology, with a brief treatment of materials and processes, rock types, stratigraphic principles, structure and tectonics, land forms and geologic time in the first few chapters. Subsequent chapters discuss the geologic character of Arizona and its geologic history. Several special topics will be discussed, including volcanism, Meteor Crater, economic and environmental geology, because they appear to be of nearly universal interest to state residents and visitors. Numerous photographs and graphic illustrations have been included to illustrate some of the more interesting, beautiful, or important geologic features.

ACKNOWLEDGMENTS

We wish to acknowledge the contributions of the many geologists who have added their observations and interpretations of Arizona geology to the general store of knowledge in the form of many publications. Those geologists are too numerous to mention, but the history of such contributions begins with the earliest European and Mexican explorers who traversed the region and recorded their observations, and is still in progress. Those whom we have relied upon most heavily are acknowledged, along with their published work, as references at the end of the book. We are especially indebted to the following individuals for their technical assistance in the production of this book: Connie Erickson for typing the manuscript; James Q. Brown for photographic assistance; Patricia Haggerty and Sue Selkirk for drawing several illustrations; and Michael Sheridan, Carleton Moore and Terry Danielson for reviewing portions of the manuscript. Special thanks go to Katherine Kron for thoroughly editing the first edition of this book, and for her many helpful suggestions for improvements.

Several of our colleagues, including Troy Péwé, Wesley Peirce, Terah Smiley, Stanley Beus, Ron Blakey, David Brumbaugh, Steven Reynolds, and Clay Conway have shared their knowledge and allowed us to use some published illustrations of their research. We express our appreciation to all these people for their contributions to this book, however, we alone are responsible for the accuracy of this work and for any omissions.

The appearance of this book is enhanced by the cover painting by Cynthia Bennett, whose creative talent captures the impressions of Arizona's magnificent landscapes. Her generous contribution is greatly appreciated.

We also appreciate the availability of facilities and services of Northern Arizona University, Arizona State University, the Arizona Bureau of Geology and Mineral Technology, the Museum of Northern Arizona and the United States Geological Survey; without which the completion of this book would have been much more difficult.

PART I

EARTH MATERIALS, PROCESSES AND GEOLOGIC PRINCIPLES

The evolution of the earth's crust and landscape has been a long and complex process. The sequence of events has varied from one place to another, resulting in combinations of rock units, tectonic events and landforms that are unique to each portion of the earth. For example, Arizona is characterized by rocks, structures and landscapes that are different from those found in New Mexico, Colorado, Utah or California. Even if some geologic features should appear similar in other areas, it is likely that they were formed at times different from those in Arizona. This uniqueness of the geologic character of Arizona is the result of the cumulative effects of the nature and timing of geologic processes such as tectonism, plutonism, erosion, sedimentation, and volcanism which have operated over a very long time. The rocks of the earth's crust contain numerous clues to their origin and to subsequent events that have affected them. These clues can be observed and their significance interpreted by the application of a few relatively simple geologic principles and techniques which have been thoroughly tested by several generations of geologists and other natural scientists. Chapters one through eight discuss briefly the nature of earth materials and the geologic principles and concepts that are necessary to understand and interpret the geologic history and evolution of Arizona.

ROCKS AND MINERALS

Since the beginnings of the planet more than 4.5 billion years ago it has been continually changing. Mountains have arisen and been leveled to rolling plains, ocean basins have disappeared and others have been created, the continents have shifted their positions and climates have changed. The earth has evolved through episodes of geological activity repeated numerous times at different places around the globe. Each episode has left its mark by the creation of new rocks or the transformation of older ones. In a sense the earth has aged, with new wrinkles being added with the passing years. For the most part our knowledge of the events in earth history comes from the study of the rocks exposed at the surface and from a careful examination of the scenery. Few places on earth display such a variety of geological processes in such a spectacular fashion as Arizona does. Located in an arid region, the state is but sparsely covered with vegetation and the rocks in many places are laid bare for the beholder.

To understand the geological history of Arizona it is necessary first to understand the basic rock types and the processes responsible for their formation. Although each region on earth has a unique history, the processes at work are everywhere similar. The first portion of the book introduces basic geological principles illustrated with Arizona examples, but this information may be applied anywhere one's travels may take him, where rocks are exposed at the surface.

All rocks can be grouped into one of three basic types, each of which is created by geological processes acting on one of the other rock types. These transformations of one rock type to another have recurred many times during the earth's history. They may be conceptualized using a diagram called the *Rock Cycle* which illustrates all possible rock changes (Figure 1–1).

Before explaining this, it is necessary to realize that all rocks, regardless of type, are composed of *minerals* (except for coal and a few volcanic glasses), and that the minerals themselves are composed of orderly arrays of atoms with specific crystal structures and chemical compositions. Since the processes that convert one rock type to another take place at the mineralogical or atomic level it is necessary at the beginning of the book to consider minerals and a little chemistry, as a foundation for all that follows.

Since all rocks are derived from one of the other types, and since the first rocks on earth have long ago been converted to something else, where we begin in the Rock Cycle is arbitrary. The most fundamental rock type is *igneous rock*, which forms when minerals crystallize from melted rock, or *magma*. Magma is a hot liquid which forms somewhere at depth from the melting of other rock. Being liquid, magma will move up into cooler rock or reach the surface and erupt as a volcano. If the magma cools and crystallizes before it reaches the surface the resulting igneous rock is said to be *intrusive*, whereas when magma pours out at the surface the igneous rock is *extrusive*.

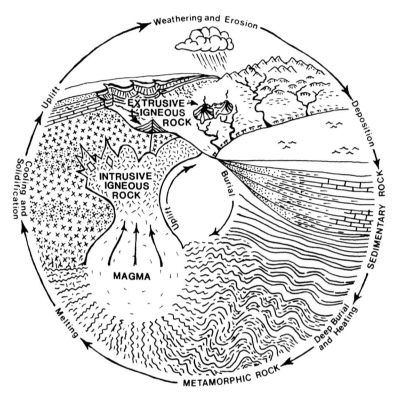

Figure 1-1. Rock cycle.

At the earth's surface the processes of weathering and erosion continually wear away whatever rocks are exposed. They are broken down into individual mineral grains or rock fragments and transported by currents of water and wind to places where they are deposited as *sediments.* Igneous rocks which form at depth eventually reach the surface as they are uplifted and the overlying rocks are removed by erosion. As sediments accumulate, the bottom of the pile is subjected to increased pressure and temperature which causes compaction and eventually *lithification,* the process of transferring loose sediment into solid *sedimentary rock* by the cementing together of the individual grains.

If sedimentary rocks are buried to great depths and subjected to high temperatures and pressures, new mineral grains will grow at the expense of others. In essence, the atoms of a mineral are rearranged into another combination in the rock. This process, which takes place in the solid state, is called *recrystallization* or *metamorphism,* and the resulting rocks are *metamorphic.*

If metamorphic rocks are subjected to high enough temperatures they will begin to melt. The magma upon cooling will produce igneous rock, thus completing the rock cycle. It should be noted that uplift may occur at any stage in the cycle and that sedimentary, metamorphic and igneous rocks may be weathered to produce sediment for more sedimentary rocks. The reasons for uplift are varied and will be discussed later in the book.

For many people the first contact with geology comes through an interest in minerals, whether it be as a museum visitor or as a "rock hound" building a collection first hand in the field or by scouring the mineral and rock shops. The beautiful, colored specimens faceted with crystal faces are rare in nature. Usually crystals grow tightly together interfering with each other so that they neither grow to a large size nor develop smooth crystal faces.

By definition a mineral is a naturally occurring element or compound which has a specific chemical composition *and* crystal structure. Minerals are formed in nature, not in the laboratory, distinguishing themselves from the many substances created by the chemist.

Although more than 90 different elements are known to occur on earth, only eight of them make up more than 98% of the earth's crust (see Table 1–1). The most abundant is oxygen. Though most of us think of oxygen only as a gas necessary to life, in fact it readily combines with most other elements and is found in most of the different mineral types.

The second most abundant element is silicon, and the most abundant group of minerals are the *silicates,* combinations of silicon, oxygen and one or more of the other abundant elements. The basic building block of all the silicates is the *silica tetrahedron,* four atoms of oxygen surrounding an atom of silicon (Figure 1–2). The radius of the oxygen atom is large compared to the silicon atom and the oxygen atoms surround the silicon atom tightly. The geometrical figure produced by connecting the centers of each of the oxygen atoms is four sided with triangles on each side—the tetrahedron.

Silica tetrahedra may be combined in various ways by the sharing of an oxygen atom of one tetrahedron with one or more adjacent tetrahedra. The silicate minerals are combinations of the other most abundant elements with one or another of the possible silica structures.

Table 1–2 shows the most common of the silica structures and the minerals which they form. The chemical formula in the table tells the ratio of one element to another in a given mineral. The symbols are abbreviations for the different elements; the numbers following the symbols indicate the relative abundance of that element in the mineral. No number means "one." In quartz, for instance, for every silicon atom there are two oxygen atoms. In olivine the silica tetrahedron combines with iron and magnesium (Table 1–3). The parentheses in the formula mean that for every SiO_4 there are two atoms of either iron or magnesium or both in some ratio. Most commonly

TABLE 1–1.   Most Abundant Elements in Earth's Crust

Element	Chemical Symbol	Abundance in Weight %
Oxygen	O	46.6
Silicon	Si	27.7
Aluminum	Al	8.1
Iron	Fe	5.0
Calcium	Ca	3.6
Sodium	Na	2.8
Potassium	K	2.6
Magnesium	Mg	2.1
All other elements		1.5

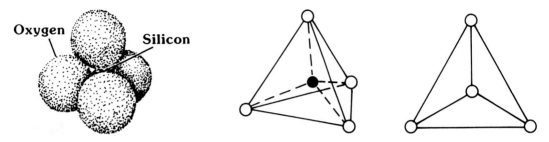

Figure 1-2. Silica tetrahedron, 4 atoms of oxygen surrounding an atom of silicon. Lines connecting the centers of the oxygen atoms create the tetrahedron, a geometrical figure with four triangular faces.

both iron and magnesium occur, with their ratio being dependent on their relative abundances in the magma from which the olivine crystallized. Iron and magnesium may occupy the same sites in the olivine structure because they are about the same size.

The atoms in a mineral are held together by attractive forces known as *bonds*. The strength of bonding in a particular mineral determines its hardness.

As a crystal grows, the atoms arrange themselves in orderly, repeated patterns characteristic of the particular mineral. If the mineral is crystallizing in a magma it will not be constrained by adjacent minerals and smooth, outer surfaces or faces will grow, reflecting the structure of the atoms that are combining. Because of the unique crystal structure, the crystal faces on one mineral type will always be found to make characteristic angles between the faces. A classification scheme has been devised which groups all the different minerals into one of six possible geometries, with a wide variety of possible variations within the six groups. If one is interested in the crystallography of minerals, information is available in most mineral handbooks.

The crystal form of some of the silicates can be related to the structure of the silica tetrahedra. For instance the chain silicates, pyroxene and amphibole, are prisms, elongated parallel to the chains. The micas are sheetlike and hexagonal, as are their crystal structures.

The way that one identifies a mineral that he finds in the field is to consider its physical properties, the way it looks, its color, its luster, the angles between crystal faces if they exist. Each mineral has a characteristic set of physical properties. A single property, such as color, is not good enough usually by itself, for many minerals are the same color. One needs to consider several properties to narrow the choices.

One important physical property is *hardness*. The hardness of a mineral can be determined by comparing it to a reference scale called the Moh's Hardness Scale. It includes 10 minerals numbered from 1 to 10 in order of increasing hardness (Table 1-4). The hardness scale includes the softest known mineral, talc, and the hardest, diamond. To tell the hardness of an unknown specimen one would rub it against the minerals in the reference set and see which scratches which, the scratched piece obviously being the softer.

TABLE 1–2.   Structures of Silicate Minerals. Based on Combinations of the Silica Tetrahedon.

	Structure	Formula of Silicon-Oxygen Unit	Examples
	Single Tetrahedron	(SiO_4)	Olivine
	Chains	(SiO_3)	Pyroxene: augite
	Double Chains	(Si_4O_{11})	Amphibole: hornblende
	Sheets	(Si_2O_5)	Mica: muscovite biotite
	Three Dimensional Framework	(SiO_2)	Quartz Feldspars

TABLE 1-3.   Some Common Minerals and Their Physical Properties

Mineral	Chemical Composition	Common Colors	Luster	Hardness	Cleavage
Quartz	SiO_2	Clear colorless, milky white, pink, gray, etc.	Glassy	7	None, concoidal fracture
Plagioclase	$NaAlSi_3O_8$ $CaAl_2Si_2O_8$	White, gray, colorless	Glassy to Pearly	6	Two directions, intersect at about 90°
Orthoclase (K-feldspar)	$KAlSi_3O_8$	Pink, white, gray, colorless	Glassy	6	Two directions, about 90°
Muscovite (white mica)	$KAl_3Si_3O_{10}(OH)_2$	Clear to silvery green, or yellow	Silky or Pearly	2-2.5	One direction, cleaves to thin sheets
Biotite (black mica)	$K(Mg,Fe)_3AlSi_3O_{10}(OH)_2$	Dark brown to black	Silky or Pearly	2.5-3	One direction, cleaves to thin sheets
Hornblende (Amphibole Group)	$Ca_2Na(Mg,Fe)_4(Al,Fe)(Al,Si)_8$ $O_{22}(OH)_2$	Dark green to black	Glassy	5-6	Two directions, intersect at 56° and 124°
Augite (Pyroxene Group)	$Ca(Mg,Fe,Al)(Si,Al)_2O_6$	Dark green to black	Glassy	5-6	Two directions, intersect at 90°
Olivine	$(Mg,Fe)_2SiO_4$	Green to brown	Glassy	6.5-7	None, conchoidal fracture
Calcite	$CaCO_3$	Colorless to white	Glassy to Earthy	3	Three directions, intersecting at 75° and 105°
Pyrite	FeS_2	Pale brass-yellow	Metallic	6-6.5	none
Chalcopyrite	$CaFeS_2$	Brass yellow	Metallic	3.5-4	none

TABLE 1–4. Moh's Hardness Scale.

10	Diamond	
9	Corundum	
8	Topaz	
7	Quartz	
6	Orthoclase feldspar	increasing hardness ↑
5	Apatite	
4	Fluorite	
3	Calcite	
2	Gypsum	
1	Talc	

A property possessed by some minerals is *cleavage,* the characteristic of breaking or fracturing along planes. The cleavage planes follow planes in the crystal structure where bonding is weaker than in other directions. Cleavage may occur in one, two or three directions. Mica has extremely well developed cleavage in one direction, parallel to the sheets in the crystal structure. Pyroxene and amphibole each have two directions of cleavage cutting between the chains in their structure. In pyroxene the planes are about perpendicular to each other, whereas in amphibole they intersect at about 60° angles. Calcite is a common mineral which has three good cleavage directions, producing faces that intersect at 75° and 105°. Sometimes crystal faces are mistaken for cleavage faces. Remember that the former are due to growth of a crystal, while the latter are due to breakage.

The nonplanar fracture of some minerals is characteristic, particularly if it is conchoidal, as occurs when glass is broken. Quartz is a mineral that often fractures conchoidally.

Another physical property is *luster,* the way in which a mineral reflects light. The two main types are metallic and nonmetallic. An important group of metallic minerals are the sulfides, many of which are mined as ores. Most minerals, however, are nonmetallic. Of these the most common luster is vitreous or glassy. But in addition, some minerals are dull and earthy, and others may be described as resinous or pearly.

Finally, there is the physical property of *specific gravity,* or the heaviness of a mineral. The value is the ratio of the weight of a mineral compared to the weight of an equal volume of water. Special devices are used for this determination, but for field identification it is helpful simply to heft the specimen to feel whether it is *relatively* heavy or not. Most minerals have a specific gravity of about 2.5, (two and a half times denser than water), but some are greater than 7 or 8.

The most important rock-forming minerals and several of the common ore minerals are listed in Table 1–3. Quartz is the most familiar mineral to many people and is common in each of the three rock groups. It is usually transparent or transluscent, clear to gray in hand-specimen, but colorful varieties occur as well; for example, pink—rose quartz, gray—smoky quartz, purple—amethyst. Milky white quartz is a common variety in veins of igneous rock.

The most abundant minerals in the earth's crust are the feldspars, which come in two varieties: orthoclase or microcline feldspar containing potassium, and plagioclase feldspar containing a mixture of sodium and calcium. The two types are sometimes difficult to tell apart in the field.

Plagioclase is always white and orthoclase may be white, but sometimes it is pink, a color that permits positive identification.

The micas are distinctive because of their sheetlike habit. Two varieties are distinguished on the basis of color: muscovite is white or silvery, whereas biotite is black or very dark brown.

Calcite, a carbonate mineral, is the most common nonsilicate. It is found in the sedimentary rock limestone, but there the individual crystals are generally too small to be seen with the unaided eye. Good calcite crystals may be found growing in open spaces or vugs in rocks, often in openings in certain volcanic rocks.

Pyrite and chalcopyrite are the most common sulfide minerals. Both are a metallic gold color, with the chalcopyrite a slightly darker yellow. Since these minerals are sometimes mistaken for the valuable metal they are called colloquially "fool's gold." Chalcopyrite contains copper and is an important ore mineral in Arizona.

The mineral most associated with Arizona is turquoise, the bluish green, semiprecious stone commonly used in Navajo jewelry. It is a complex copper phosphate $CuAl_6(PO_4)_4(OH)_8 \cdot 5H_2O$ associated with a number of the copper deposits in the state, where it has formed as a secondary mineral after the weathering of primary minerals in the deposit (see Chapter 12).

Green malachite and blue azurite are two copper carbonate minerals $Cu_2CO_3(OH)_2$ and $Cu_3(CO_3)_2(OH)_2$ which also formed as secondary products of the weathering and reprecipitation of copper ores. The mines around Bisbee have produced crystalline and banded specimens which grace museum collections around the world.

Peridot, the gem variety of olivine, occurs in an exceptional deposit on Peridot Mesa near San Carlos, where nodules of the olivine have been brought to the surface in lavas generated at depth beneath the crust of the earth.

A great deal of pleasure can be derived from pursuit and collection of minerals in the field, both in the exploring of new territory and the excitement of discovery. But minerals also continue to provide enjoyment as objects of beauty when displayed back home. For a complete listing of Arizona's minerals, with many colorful photographs, "Mineralogy of Arizona" by Anthony, Williams and Bideaux is recommended.

CHAPTER 2

WEATHERING AND EROSION

The earth is everywhere in a state of decay. Rocks are constantly being broken down into smaller particles of rocks or minerals, which in turn are reduced into still smaller particles. As they are broken down, they are moved by gravity, running water, wind or ice, from high regions into low regions. The processes by which rocks are broken down and the particles moved are called *weathering* and *erosion,* and the net effect is a gradation of the earth's surface—the reduction of high regions and the filling in of low regions.

Weathering is the decomposition of solid materials in the earth's crust, and is accomplished through chemical, physical and/or biological activities. *Chemical weathering* is the breakdown of rocks and minerals through chemical reactions, generally between water or weak acids and the chemical elements such as iron, calcium, and sodium that are primary constituents of minerals. Weak solutions of hydrochloric, sulfuric and nitric acids are formed naturally by reactions in the atmosphere or hydrosphere, and they affect the minerals much more rapidly than water alone. These acids form as water falls as rain through the air, with its dissolved gases, and as it flows over or through the rocks or soil of the surface. The effects of chemical weathering may be very subtle or spectacular. More subtle changes can be seen in the general softening of the rock as water seeps into myriad tiny cracks and fractures in the rock causing chemical decomposition of the minerals. The mineral particles then become separated from each other and a gravelly or sandy mantle of loose material (grus) is left as a residue on the surface, such as is found in areas of granitic bedrock like the Payson or Prescott areas (Figure 2–1, a, b, c, d). More spectacular changes may be seen in limestone exposures where the rock is dissolved and carried away leaving fissures and caverns large enough to walk into. Examples of such caverns are common in the limestone terrain of the Colorado Plateau, for example, Grand Canyon Caverns on Highway 66, 8 miles east of Peach Springs; Redwall Cavern and Rampart Cave in the Grand Canyon; and several caverns in the Tucson area such as Onyx Cave and Peppersauce Cave (Figure 2–2).

Physical weathering is the process of breaking down rock and mineral particles due to the expansion and contraction with daily or seasonal temperature changes, *freezing of water in cracks,* or breaking as a result of rocks rolling or falling down hill under the force of gravity.

Biological weathering includes the reduction of rocks and minerals by root wedging, burrowing or trampling by animals, or ingestion of minerals in soil or sediment by burrowing animals. The breakdown of rocks, soil and sediment of the earth's surface by human activities such as mining and construction excavation could be considered mechanical weathering. The primary driving forces for weathering are solar heating and gravity.

Figure 2–1,a. Outcrop of weathered granitic bedrock near Prescott.

Figure 2–1,b. Spheroidal weathering of granite. Note solid core surrounded by decomposing granite.

Figure 2–1,c. Closeup of differentially weathering minerals in granite porphyry. Phenocrysts of K-feldspar, groundmass crystals of quartz, feldspar and biotite. Biotite weathers most strongly.

Figure 2–1,d. Granite grus, consisting of mineral grains, and rock fragments removed from source of weathering.

Figure 2-2. Solution cavern in Redwall Limestone along Colorado River in Marble Gorge.

Erosion is the movement of weathered rock and mineral material from one place to another on the earth's surface, generally from higher to lower places. The primary agent of erosion is running water in response to gravity; however in some areas wind, glaciers, or gravity alone may contribute greatly to erosion (Figure 2-3).

Running water moves loose rock and mineral material down slopes first by sheet wash, then collects into small rivulets which converge into larger and larger streams. Wherever the water flows it carries suspended or dissolved minerals as long as the water velocity is great enough. When water velocity decreases, the *load* of sediment is deposited forming layers of gravel, sand silt, clay, or mineral precipitates (Figure 2-4). These layers of sediments, if left undisturbed and protected by other layers of sediments, are the materials from which sedimentary rocks are formed.

Figure 2-3. Boulders in Oak Creek Canyon transported by waters of Oak Creek during flood stage.

Figure 2-4. Gravel and sand layers deposited by stream action during the Pleistocene by the Little Colorado River, Grand Falls area.

CHAPTER 3

STRATIGRAPHIC PRINCIPLES

Most of our knowledge of the history of the earth has come from the direct observation and interpretation of clues about past events that are contained in the rocks of the earth's crust. Such clues may be found in the mineral composition, texture, sedimentary structure or fossil content of the rock, which tell us something about the environmental conditions that existed at the time the sediments were laid down or, in the case of igneous or metamorphic rocks, at the time when the magma cooled and crystallized or the metamorphism occurred. The significance of sedimentary features is discussed in Chapter 5 and of igneous and metamorphic features in Chapter 6.

During the past two hundred years, several basic principles of geology have been developed and tested, and are now the basis on which nearly all geologic interpretation is done. These basic principles are summarized as follows.

Uniformitarianism—The geologic features of the earth's surface and crust could have been formed by the processes of erosion, deposition, volcanic activity and crustal movements that are presently active on or in the earth's crust. Sometimes paraphrased as "the present is the key to the past," this principle is fundamental to the interpretation of paleoecology and environments of deposition.

Original Horizontality—Sediments being deposited in a basin of deposition (lake, ocean, etc.) will accumulate in stratified layers in an essentially horizontal attitude. The principle is fundamental to the recognition of structural deformation of the crust, and the time of its occurrence.

Superposition of Strata—In an undeformed sequence of strata, the oldest (first formed) layer will be the lowest, and each consecutive layer will be younger than the one below it. This principle is fundamental to the recognition of the relative age of strata and the interpretation of geologic time.

Cross-cutting Relationships—Any geologic feature that cuts across or is intrusive into a body of rock, must be younger than that body of rock. This principle is fundamental to the recognition of the age of faults, folds and igneous intrusions.

Inclusions—Any body of rock which contains fragments of other rocks must be younger than the rocks from which those fragments were derived. This principle is fundamental to the interpretation of the age of conglomerates, and sources of sediments, time of erosion and relative age of intrusive rocks.

Fossil Succession—Groups of fossil plants and animals succeed each other in a stratigraphic sequence, in a definite and nonrepetitive order. This principle is fundamental to the subdivision of the stratigraphic sequence into intervals of geologic time.

These few principles provide a basic frame of reference within which most geologic problems involving the history of the earth can be worked out. They are used daily by geologists around the world.

Stratigraphy

Much of Arizona's land surface is composed of stratified (or layered) rocks, either sedimentary or volcanic. These rocks are hundreds to thousands of feet thick and sometimes contain ground water, oil and gas, and certain mineral deposits. Those stratified rocks also contain clues as to the geologic history of Arizona. For these reasons we are interested in learning more about them, and to do so, we must apply the basic principles discussed above, plus some special techniques of a science called *stratigraphy*.

Stratigraphy is the branch of geology which deals specifically with the origin, sequence, composition and relationships of stratified rocks. Stratigraphic studies are normally applied to sedimentary rocks; however, lava flows or ash falls of volcanic origin may be included within sedimentary sequences, and may be examined by similar techniques.

The first concern of stratigraphy is with the origin of sedimentary rocks, that is, the processes by which the sediments were transported and deposited and the nature of the basin in which they were deposited. Obviously sediments are derived from elevated land surfaces by processes of weathering and erosion and are transported downhill to low elevations on the Earth's surface. Tectonic forces cause the uplift or subsidence of the earth's crust, and such movements have affected all parts of the earth's surface at different times. Any given area of the Earth's continents has during some time in its history been uplifted above sea level and subjected to erosion, like Arizona is today. However, at other times, that same area of the Earth's surface may have subsided below sea level and, instead of being eroded, been covered by layer upon layer of sediment on the bottom of a shallow sea. The subject of tectonics will be discussed thoroughly in Chapter 7, so for the purposes of this discussion of stratigraphic principles let us say that throughout most of the two billion years of earth history detectable in Arizona the state has been located near the eastern margin of a broad region of subsiding crust that was periodically flooded by shallow seas from the west (the Cordilleran Geosyncline), from the south (the Sonoran Geosyncline), or from the east (Western Interior Seaway). During the times of subsidence from two billion years ago to 250 million years ago, marine sandstones, shales and limestones accumulated as broad sheets of sediments across most of the state. Incorporated in those sediments were the skeletal remains, tracks and burrows of marine organisms such as trilobites, brachiopods, corals, mollusks, echinoderms and foraminifera.

Facies

Since the nature of sediments is due to the environmental conditions existing where they are deposited, we would not expect the sediments deposited everywhere during a given time to be the same. For example, the sediments accumulating on the sea floor of a broad continental shelf will normally range from river delta and beach sands near shore where wave and current action is strong enough to move sand, to mud and clay further offshore beyond strong current and wave action. Even further offshore, beyond the reach of clastic sediments eroded from land, only the sediments produced in place, by biological or chemical sedimentation, are formed. These are usually limestones. Where such variety of sediments occur at the same time, they will grade laterally into each other. Such lateral gradation of sediments (or sedimentary rocks) is called sedimentary (or stratigraphic) *facies* (Figure 3–1). The Cambrian rocks in the Grand Canyon provide a classic example of facies, where the Tapeats Sandstone is the near-shore sand facies, the Bright Angel Shale is the intermediate facies formed farther offshore, and the Muav Limestone

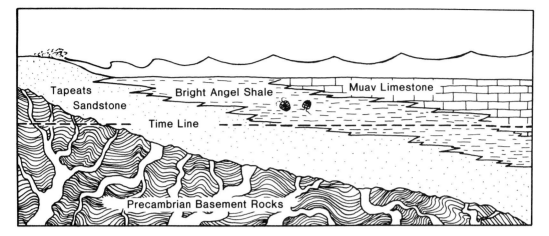

Figure 3–1. Diagrammatic cross-section of Cambrian rocks in Grand Canyon illustrating lithofacies formed during a marine transgression over an erosional surface on Precambrian plutonic rocks.

is the outer shelf deposit of carbonate muds formed beyond the reach of clastic sedimentation during the Cambrian Period. Such a lateral variation in sedimentary rocks is called, more specifically, *lithofacies*. Lateral variations in fossil content also occur due to the adaptation of marine communities to varying conditions within the depositional basin. Such variations occur in modern environments and, from the fossil record, we know they also happened in the past. Such lateral variations of fossil content in rock units is called *biofacies*. If we can determine the position of certain types of fossils relative to the shoreline, ocean depth, salinity or other environmental variant, then those types of fossils may be used as *ecologic* (or paleoecologic) *indicators* for the rocks in which they are found. A good example of biofacies can be observed in the Cretaceous rocks of Black Mesa where there is a lateral variation from terrestrial coal deposits to nearshore or lagoonal oyster reefs in the Dakota Sandstone to offshore marine ammonites, clams, snails, corals, sharks, and planktonic foraminifera in the Mancos Shale (Figure 3–2).

Unconformities

In the past there were intermittent periods when part or all of the state was raised above sea level resulting in the accumulation of nonmarine sediments on low lying coastal plains, deltas, or stream flood plains. During times when the area was raised higher, erosion of some of the previously deposited sediments occurred, thereby removing some of the record of geologic history of the state. There were several such periods of uplift and erosion during the Precambrian and Paleozoic history of Arizona, most notably during the Proterozoic (about 800 million and 1.7 billion years ago), the Ordovician and Silurian (500 to 400 million years ago), the late Devonian (375 million years ago), the late Mississippian (300 million years ago) and late Permian (250 million years ago), after which time the region has remained generally above sea level.

The results of such uplifts and erosional events has been a series of gaps in our knowledge of geologic history caused by the removal of rocks of certain periods of time. The result is much as

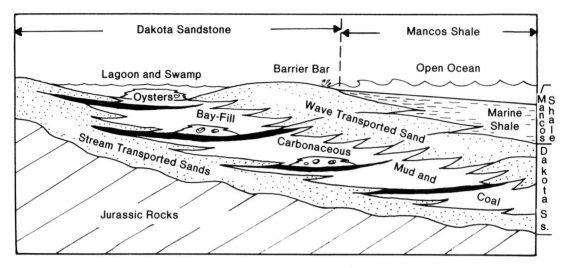

Figure 3-2. Diagrammatic cross-section of Cretaceous rocks in Black Mesa illustrating lithofacies and biofacies formed during marine transgression over an erosional surface on Jurassic sedimentary rocks. The depositional environments of the Dakota Sandstone and Mancos Shale are depicted across the top of the diagram.

if a few pages were torn out of a very detailed history book. Such gaps in the geologic record are called *unconformities.* Most unconformities are recognizable in the field due to the evidence of structural discordance and erosional features between rock units; however, in some cases there may have been no structural disturbance or pronounced erosion. In such cases the missing record can be detected by the absence of fossils characteristic of certain time intervals in adjacent regions, such as California and Nevada.

Three basic kinds of unconformities are recognized, all of which are represented in the rocks of the Grand Canyon and other parts of the state. *Nonconformities* are unconformities in which the rocks beneath the erosional surface are plutonic or metamorphic, and rocks overlying the erosion surface are sedimentary (Figures 3–3, 5–5 and 10–1). Nonconformities indicate that a long period of time is not represented, during which major deformation of the older rocks occurred, probably including major *orogenic* (mountain building) *uplift,* followed by erosion of the mountain range down to a near level surface, then subsidence to below sea level. Such an event occurred throughout Arizona between the formation of the older Precambrian igneous and metamorphic rocks (Vishnu Schist, Pinal Schist, etc.) and the deposition of basal younger Precambrian rocks (Bass Limestone, Dripping Spring Quartzite, etc.).

Angular unconformities are those in which the rocks below the erosional surface were tilted by folding and faulting followed by their erosion to a relatively flat surface, then subsidence below sea level and deposition of horizontal sediments over the truncated edges of the eroded strata (Figure 3–4). This kind of unconformity is well developed in the Grand Canyon between the younger Precambrian Grand Canyon Supergroup (Bass, Hakatai, Shinumo, etc.) and the Cambrian Tapeats Sandstone or the Bright Angel Shale.

The third kind of unconformity, called a *disconformity,* is one in which the strata above and below the erosional surface are parallel. There may or may not be a prominent erosional surface, but based on fossil or other age indicators, a definite time interval is not represented (Figure 3–5).

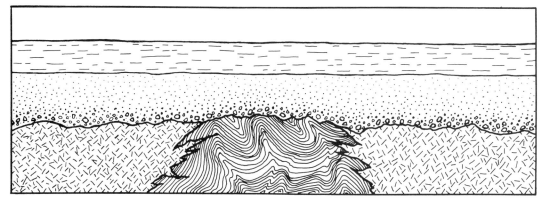

Figure 3-3. Diagrammatic cross-section of a nonconformity consisting of sedimentary rocks deposited over an erosional surface on plutonic igneous and metamorphic rocks.

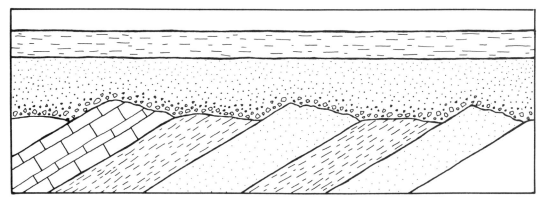

Figure 3-4. Diagrammatic cross-section of an angular unconformity consisting of sedimentary rocks deposited over an erosional surface on tilted sedimentary rocks.

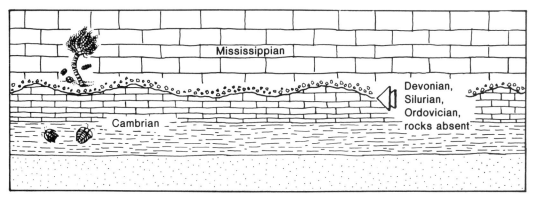

Figure 3-5. Diagrammatic cross-section of a disconformity consisting of sedimentary rocks deposited over an erosional surface on sedimentary rocks with no angular discordance, but with a significant age difference between the rock units.

This kind of unconformity is well developed at the base of the Mississippian Redwall Limestone in eastern Grand Canyon where the underlying rock is usually the Cambrian Muav Limestone. In western Grand Canyon the major disconformity lies between the Devonian Temple Butte Limestone and the underlying Cambrian Muav Limestone, with no record of Ordovician or Silurian rocks.

Correlation

Our interpretation of geologic history and the solution of many other geologic problems are dependent upon the process of equating rock units in one area with those of other areas. This process is called *correlation* and it may be based upon several kinds of evidence including similar lithology, position in stratigraphic sequence, fossils, radioactive minerals and geomagnetic reversals. Another tool for correlating past events that may be used is the presence of unconformities. They may be only of local significance or may mark a time of erosion of a whole continent. In any event they indicate a time of erosion in the stratigraphic column which may be useful in correlation.

Lithologic correlation consists of tracing the same layer of rock across country by visual observation. The criterion is lithologic similarity, therefore any gross change in rock type due to facies change would limit the application of this technique. If the lithology does not change laterally, the technique is applicable throughout the extent of the depositional basin, unless the rock unit has been partially removed by erosion. Many rock units can be correlated over long distances; for example, the Permian Kaibab Limestone can be traced continuously from an erosional and/ or depositional edge near Holbrook, westward to Las Vegas, Nevada, and from the Mogollon Rim northward (in the subsurface) into southern Utah. Throughout that area of several thousand square miles, the Kaibab Limestone maintains a general lithologic uniformity, therefore can be traced on that basis.

Other rock units in Arizona are recognizable on lithologic character, are laterally continuous and have clearly definable lower and upper boundaries. Such a rock unit has been given the formal designation of *formation* which is the basic unit of geologic mapping. A few examples of such formation units are: Coconino Sandstone, Naco Formation, Tapeats Sandstone, Redwall Limestone, and so forth. Such mappable rock units are conventionally named after some geographic feature or some characteristic of the rock unit in the type area, or the area where the formation was described and published. The specific measured section upon which the published description was based is designated its *type section*. The type section of the Kaibab Limestone is in Kaibab Gulch, north of Grand Canyon; the type section of the Tapeats Sandstone is in Tapeats Creek in the Grand Canyon; and the type section of the Naco Formation is in the Naco Hills near Bisbee. The Redwall Limestone type section is in Redwall Canyon in the Grand Canyon, where it was named by Gilbert in 1875 for its red surface color and cliff-forming character.

Formations may be combined into larger mappable units called *groups*. Several examples of groups are: Tonto Group (Tapeats, Bright Angel and Muav formations); Unkar Group (Bass, Hakatai, Shinumo, Dox and Cardenas formations); Glen Canyon Group (Wingate, Moenave, Kayenta and Navajo formations), and others. In some cases groups may be formed by subdividing a formation into more than one designated new formation, thus necessitating the "raising" of the previous formation to group status. An example of this occurred in 1975 when E. D. McKee subdivided the Supai Formation of the Grand Canyon into four formations and raised its status to the Supai Group.

Formations may also be subdivided into formally designated units called *members,* or *tongues.* Examples are the four members of the Redwall Limestone that are based on mappable minor variations in the lithologic character of the formation. Where formations are intergradational, the persistence of a small part of one unit into the other is called a tongue. For example the Boucher Tongue of the Muav Limestone extends eastward from the Bass Trail to the Bright Angel Trail as a mappable dolomite unit in the Bright Angel Shale.

Formations are strictly rock units and a given rock type (e.g., sandstone) may have formed at any time in the geologic past, from Precambrian to Cenozoic time. Rock units alone contain no evidence of their geologic age, and a formation may even be of different ages in different places. Many rock units were laid down as sediments during a marine transgression or regression therefore must be of different ages in different places (e.g., the Tapeats Sandstone).

The correlation of a laterally variable formation may be facilitated if it is immediately above, below or between laterally continuous formations. Such correlation of a rock unit is done by the *position in a stratigraphic sequence* and is often a useful technique. For example the Toroweap Formation varies in lithologic character from predominantly limestone in extreme northwestern Arizona to predominantly sandstone in eastern Grand Canyon. However throughout that distance the Toroweap is overlain by the Kaibab Limestone and underlain by the Coconino Sandstone, both of which are quite uniform and readily recognizable. The rock unit between them (i.e., the Toroweap Formation) must be correlative throughout their areal extent.

Correlation by fossils is based on the observation that the fossil content of rock units varies with their stratigraphic position. Such variation is due to the fact that there were certain kinds of organisms living on the earth at any given time in the geologic past which were different from those that lived before or after that particular time (Table 4–1). Such variation in fossil content of different stratigraphic levels is an observed fact (*Principle of Fossil Succession*), and the explanation for such change is organic evolution. If two rock units were deposited in similar environments during the same period of time, the fossil content would be similar and the correlation (same age) would be established. If two rock units were deposited in different environments, for example, marine and fresh water, or during different periods of time, the fossil content would be dissimilar—in one case due to ecological factors, in the other case because organic evolution would have caused changes in fossil composition. In the case of ecologic variation, direct correlation by fossils would not be possible. However, chronologies have been established based on the stratigraphic ranges of fossils from different environments, (e.g., marine invertebrates, or land mammals or fresh water mollusks and ostracodes) therefore it is possible to correlate such dissimilar fossil assemblages by comparing each with the appropriate established chronology. If they both prove to be of the same geologic age (e.g., Devonian) then correlation is established. For example the Devonian Martin Formation in central Arizona has been divided into two members—the Jerome Member with marine fossils (Fig. 5–10) and the Becker's Butte Member with land plant fossils (Fig. 10–3). In order to correlate the two members with other Devonian formations on the basis of fossils, and determine the geologic age, each would have to be compared with different chronologies, one based on the stratigraphic ranges of marine organisms, the other based on stratigraphic ranges of land plants.

Fossil correlations, especially those based on identical species in different formations, are indicators of identical age of the formations. The reason is that any species or group of species existing together will only exist on earth for a short time in terms of geologic time. Before that time interval those species would not yet have evolved, and after that time they would have become extinct or evolved into different forms. Such short time intervals characterized by discrete and nonrepetitive aggregates of fossil species are called *time-stratigraphic zones*. Such zones are true time markers that indicate an "event in time," some of which may be as brief as 250,000 years. One such zone established by McKee and Resser (1945) in the Middle Cambrian Bright Angel Shale contains the trilobites *Alokistocare althea, Anoria tontoensis* and *Glossopleura mckeei*, and is called the *Glossopleura-Alokistocare* Zone. The zone is recognizable in numerous localities from Grand Wash Cliffs to eastern Grand Canyon, a distance of 95 miles.

Time-stratigraphic zones are grouped together into larger units called *stages*. Stages are groups of strata that are characterized by fossil assemblages that are peculiar to them, that is, they do not occur at any other level of the stratigraphic section. Stages are recognizable over greater distances than zones, and may be contained within one formation or may include several formations. For even broader correlations over wider geographic areas, stages may be grouped into *series* and series are grouped into *systems*. For example, there are three *series* (Waucoban, Albertan and St. Croixan) within the Cambrian System, which includes all strata of Cambrian age in western North America. All other geologic systems recognized around the earth may be similarly subdivided into smaller and smaller subdivisions. All such time-significant subdivisions of the stratigraphic column are called time-stratigraphic units, which are clearly separated from rock-stratigraphic units discussed above.

A third type of stratigraphic unit includes subdivisions of geologic time called time terms and are not directly related to rocks of the earth's crust. The three types of stratigraphic terms are illustrated in Table 3–1.

Another technique of correlation particularly applicable to igneous, plutonic or volcanic rocks, is based upon the measurement of radioactive minerals and their decay products to interpret the age of the rock in years before present. This technique will be discussed in Chapter 4 in some detail, so let it suffice at the moment to say that such *radiometric dates* provide another technique for correlating rock units from one area to another. The dates available on older Precambrian rock units such as the Zoroaster Granite in the Grand Canyon and the granodiorites in central Arizona indicate they were both formed at the same time, 1,700 million years ago. Similarly, radiometric dates on Cenozoic volcanic rocks across Arizona allow comparison of their times of formation; for example, the San Francisco Peaks were formed between about 1.8 and 0.2 million years ago, the White Mountains were formed between 10 and 1 million years ago. In contrast,

TABLE 3–1. Stratigraphic Terms

Time Terms	Time-Rock Terms	Rock Terms*
Era	Terrain	Group
Period	System	Formation
Epoch	Series	Member
Age	Stage	Tongue
Chron	Zone	

*Note: there is no direct equivalence between the terms in this column and the Time-Rock terms.

most of the volcanic rocks in southern Arizona are considerably older, about 30 to 15 million years old. The radiometric dating of geologically young (100 to 70,000 years old) sediments containing fossils is possible by the use of radiocarbon dating of wood, bone or shell material. This technique is used for correlating surficial deposits of stream or lake origin in Black Mesa and some basins in southern and central Arizona, and is especially useful in the interpretation of the timing of late Cenozoic climatic changes.

A recently developed technique of correlating sedimentary and volcanic rocks based upon *geomagnetic reversal chronology* is adding to our knowledge of the age relations of rock units across the state. It has been applied particularly to the correlation of Cenozoic nonmarine strata in southern and central Arizona, with rocks of similar age in other areas including the standard Cenozoic section best preserved in deep ocean sediments. The technique will be explained more thoroughly in Chapter 4.

CHAPTER **4**

GEOLOGIC TIME

Humans are very conscious of time. We organize our activities by reference to minutes, hours, days, months and years. These time intervals are based upon physical constants in the solar system; that is, one year is the time required for the earth to revolve around the sun, one month is the time it takes the moon to revolve around the earth. We measure these years and months by watching and counting such celestial movements and recording them on calendars. Humans have been keeping track of time since the Aztecs and Egyptians, who recorded their observations on stone or clay tablets. More recently humans have devised instruments to measure the passage of time. Such instruments include primitive sundials, mechanical clocks and "atomic" clocks.

In the course of scientific investigation of the earth during the 17th and 18th centuries it became apparent that the earth had been in existence for a much longer time than that recorded by man. Evidence for this included the great thickness of sedimentary rocks on the earth as compared with the readily observable very slow rate of accumulation of comparable sediments; the change in animal and plant life during the time of accumulation of the above-mentioned sediments as observed in fossils in them; and evidence of long periods of erosion as observed in unconformities in the rock sequence. The early scientists such as Robert Hooke (seventeenth century), Leonardo daVinci (sixteenth century), and James Hutton (eighteenth century) interpreted such evidence as indicating a long time duration, but had no way of actually measuring it. Sir Charles Lyell estimated the time required for the formation of all sedimentary rocks since the age of dinosaurs (Mesozoic) would be about 250 million years.

The first breakthrough in understanding geologic time came in 1815 when William Smith, an English engineer, published his observation that different fossil assemblages occur in different strata, and were unique to certain portions of the stratigraphic column. Such fossil occurrences were recognized as potential data for differentiating rock of different ages, when it was applied within the context of superposition of strata. The result was that after a few years, studies of strata in Europe had yielded an extensive body of knowledge about the relative position of many distinctive fossil groups that formed a nonrepetitive succession throughout the European stratigraphic column. For example, in a section across England the stratigraphically lowest fossil-bearing strata were observed to contain trilobites, but they were not found in the upper ⅔ of the section. They were replaced up-section by other fossil groups, particularly ammonites. Still higher up the section the ammonites disappeared and the fossil assemblage was dominated by clams, snails and other fossils much like those living in the oceans today. The appearance and disappearance of such fossil types in the fossil record is attributed to evolution and extinction, and are considered events in time that can be used to subdivide the continuum of geologic time into segments. The major segments of geologic time, recognized by pronounced appearances and extinctions of a large variety of animals, are called *eras*. These were designated long ago as Precambrian (rocks essentially barren of fossils); Paleozoic (= ancient life, or fossils very different than organisms living today); Mesozoic (= middle life, or fossil assemblages with some organisms similar to modern types but many that are very different from modern forms); and Cenozoic (= recent life, or organisms similar to those living today with no pronounced differences). The eras were soon subdivided into *periods* based upon extinctions and appearances (due to evolution of new forms) of fossil groups within the assemblages characteristic of eras. For example, a variety of fossil groups (including corals, bryozoans, graptolites, etc.) are not present in the stratigraphically lowest trilobite-bearing strata, but make their appearance several hundred feet above the base of the section. This appearance of new animal groups is considered an event in time and is used as the lower limit of the Ordovician Period/System. The fossiliferous rocks below that event in time are assigned to the Cambrian Period/System.

Many other extinctions and appearances of fossil types have been recognized throughout the stratigraphic column and have been interpreted as events in time to mark the boundaries of time-stratigraphic units that can be traced around the world. The result is a calendar of geologic history or *geologic time scale* based upon fossil succession, in which time-stratigraphic units ranging in magnitude from eras to zones can be recognized by the study of fossil collections. An abbreviated version of the geologic time scale is shown in Table 4–1, including some of the major events as defined by the appearance and/or extinction of certain fossil groups. All of the subdivisions of the time scale that are based on fossil succession are relative time periods; that is, they give us the age of rocks relative to (older or younger than) other rocks in the standard column. Fossils alone do not give us indications of the age of rocks measured in years. For that reason the time scale based on fossils is called a *relative time scale*.

Geologic Time Scale

The landscapes of Arizona have been formed within the last several millions of years. Many of the rocks exposed in these landscape features are, however, much older. The excavation of the Grand Canyon, for example, has occurred in a relatively short interval of geologic time, probably less than 8 million years, but the rocks exposed deep in the canyon are much older, ranging from 250 million to 2,000 million years of age. An understanding of the relative and absolute age of the rocks is essential to the interpretation of the geologic history of Arizona. A "calendar of events" in the Earth's geological history called the *Geologic Time Scale* has been compiled by geologists and other workers in allied fields such as chemistry, physics and biology. This time scale (Table 4–1) should be learned as quickly as possible because it provides the necessary frame of reference for the geologic events to be discussed in the following pages.

Three different concepts or techniques of measuring geologic time and distinguishing events in time have been devised. They are: (1) *relative time concept,* based on superposition of strata and biological evolution evident in observed changes in the fossil content of the rocks; (2) *absolute time concept,* based on the rate of decay and decay products of radioactive isotopes; and (3) *geomagnetic reversal concept;* based on variations in magnetic polarity of rocks of various ages. Each of these concepts and techniques contributes to our interpretation of geologic history of Arizona as discussed in the following pages.

TABLE 4-1. Geologic Time Scale

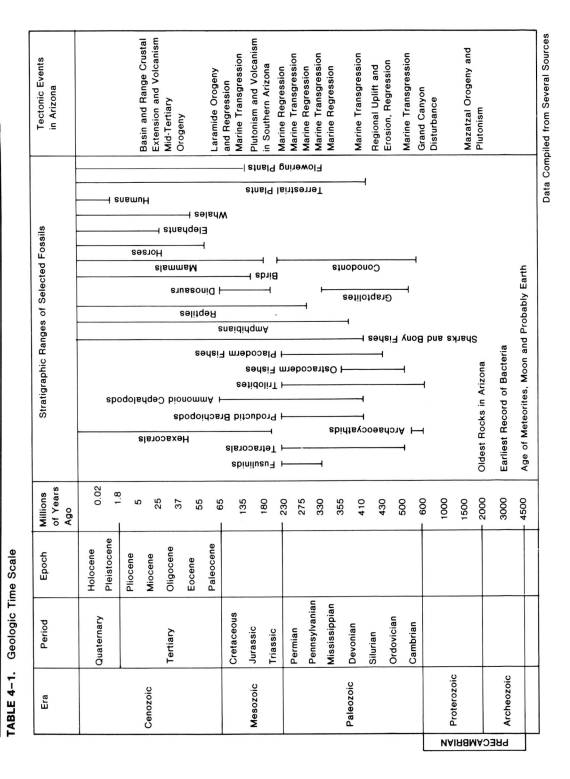

Era	Period	Epoch	Millions of Years Ago	Stratigraphic Ranges of Selected Fossils	Tectonic Events in Arizona
Cenozoic	Quaternary	Holocene	0.02		Basin and Range Crustal Extension and Volcanism
		Pleistocene	1.8		
	Tertiary	Pliocene	5		Mid-Tertiary Orogeny
		Miocene	25		
		Oligocene	37		
		Eocene	55		
		Paleocene	65		Laramide Orogeny and Regression
Mesozoic	Cretaceous		135		Marine Transgression
	Jurassic		180		Plutonism and Volcanism in Southern Arizona
	Triassic		230		Marine Regression
Paleozoic	Permian		275		Marine Transgression
	Pennsylvanian		330		Marine Regression
	Mississippian		355		Marine Transgression
	Devonian		410		Marine Regression
	Silurian		430		Marine Transgression
	Ordovician		500		Regional Uplift and Erosion, Regression
	Cambrian		600		Marine Transgression
Proterozoic			1000		Grand Canyon Disturbance
			1500		
			2000	Oldest Rocks in Arizona	Mazatzal Orogeny and Plutonism
			3000	Earliest Record of Bacteria	
Archeozoic			4500	Age of Meteorites, Moon and Probably Earth	

Selected fossils shown on the stratigraphic range chart: Flowering Plants, Terrestrial Plants, Humans, Whales, Elephants, Horses, Mammals, Birds, Dinosaurs, Reptiles, Amphibians, Sharks and Bony Fishes, Placoderm Fishes, Ostracoderm Fishes, Trilobites, Ammonoid Cephalopods, Productid Brachiopods, Archaeocyathids, Hexacorals, Tetracorals, Fusulinids, Conodonts, Graptolites.

PRECAMBRIAN

Data Compiled from Several Sources

Relative Time Scale

The general framework of the geologic time scale was developed during the nineteenth century in Europe where the larger time intervals called eras and periods were established to include the time during which discrete segments of the total stratigraphic sequence were deposited. The recognition of these time intervals is based on the observed succession of different assemblages of fossil organisms that are unique to certain segments of that total vertical sequence of sedimentary rocks. For example, the lowermost sequence of stratified rocks exposed in various parts of the earth containing essentially no animal fossils and rarely simple plants is called the Precambrian Era. Rocks of this era are found in the stratigraphically lowest igneous, metamorphic and sedimentary rocks in the Grand Canyon, the Central Mountain Province and the Basin and Range Province (Fig. 8–1). The Precambrian Era spans about 3 billion years of geologic history and is not subdivided into universally accepted time periods.

Overlying the Precambrian rocks is a vertical sequence of stratified rocks that contain abundant fossils of animals and plants that are quite different from those found in stratigraphically higher (younger) rocks. This part of the stratigraphic sequence is named the Paleozoic (ancient life) Era and is characterized by the well-known group of fossils called trilobites, along with many other extinct groups of animals and plants. The Paleozoic Era is subdivided into eight periods of time based on the appearance or extinction of different fossil groups in the stratigraphic sequence. The lower boundary is marked by the "sudden" appearance of abundant and diversified invertebrate animals and the upper boundary by the disappearance or extinction of a great variety of Paleozoic life forms. The Paleozoic Era spans the geologic time interval between 230 and 600 million years ago. Rocks of Paleozoic age in Arizona are widespread in the Plateau Province, the Central Mountain Province and in many ranges of the Basin and Range Province.

Overlying rocks of Paleozoic age is a vertical sequence of stratified rocks containing an assemblage of animal and plant fossils with a more modern aspect, especially in rocks of marine origin, although there are also some extinct major groups. This sequence has been named the Mesozoic (middle life) Era and is characterized by the presence of dinosaurs on land and a variety of extinct reptile and invertebrate groups in the oceans. Certain familiar groups of modern land animals (mammals, birds, turtles), invertebrates (crabs, certain molluscan genera) and land plants (angiosperms, cycads, pines) appeared during the Mesozoic and flourished in later time. The Mesozoic Era is subdivided into three periods based on definitive ranges of fossil groups in the stratigraphic sequence. The end or upper boundary of the Mesozoic Era in the stratigraphic sequence is marked by the extinction or disappearance of many prominent groups of land animals (dinosaurs), marine vertebrates (plesiosaurs, ichthyosaurs, and mosasaurs) and marine invertebrates (ammonites, rudistids). The Mesozoic Era spans the geologic time interval between 65 and 230 million years ago.

Rocks of Mesozoic age in Arizona are especially widespread in the eastern part of the Plateau Province in the form of flat to gently dipping or folded sedimentary formations, mostly nonmarine in origin. They are generally not present in the Central Mountain Province but are represented in the Basin and Range of the southeastern corner of the state by a thick sequence of marine sedimentary rocks and extensive intrusive and extrusive igneous rocks, and by a thick sequence of nonmarine red beds in southwestern Arizona.

The Cenozoic (recent life) Era is based upon the youngest or uppermost sequence of strata in the total stratigraphic sequence. It is characterized by a great diversification of mammals, flowering plants, and birds on land and modern invertebrates in the oceans, which has culminated in the modern fauna and flora. The Cenozoic Era is subdivided into two periods which are subdivided into seven epochs, based upon definitive ranges of genera and species, especially of mammals on land and invertebrates in the oceans. The Cenozoic Era includes the last 65 million years of geologic time up to the present. Cenozoic rocks in Arizona are predominantly volcanic, occurring commonly in all three provinces. However, Cenozoic sedimentary deposits are extensive also, occurring as thick sequences of nonmarine basin fill deposits in the Basin and Range Province and Central Mountain Region, with thin but widespread deposits on the Plateau, and deposits of marine origin in a narrow trough along the lower Colorado River.

Some of the most fascinating developments in the geologic history of Arizona have occurred during the Cenozoic Era. They will be discussed in Chapters 12, 13 and 14.

Absolute or Isotopic Time Scale

Soon after the beginning of the 20th century the discovery of radioactive isotopes in minerals and rocks led to the development of a technique to measure the age in years of rocks containing such isotopes. The principle of *isotopic age dating* is based on the process of radioactive decay of unstable elements within the crystal lattice of certain minerals. In this process a radioactive parent isotope such as uranium, changes spontaneously at a statistically constant rate to its unique daughter isotope lead, which is stable and undergoes no further change. Since the decay rate for any given pair of parent-daughter isotopes is known, and providing that the mineral has remained a close system (i.e., with no loss or gain of either isotope) the resultant ratio of parent/daughter isotopes can be used to calculate the time that the crystal and its enclosing rock matrix has been in existence. Radioactive isotopes such as uranium, thorium, potassium and others are commonly included in igneous rocks, thus such rocks are potential radioactive clocks. In Arizona, igneous rocks were formed abundantly in the Precambrian, Mesozoic and Cenozoic Eras; and improving techniques are yielding increasingly accurate records of the absolute ages of these rocks. Datable volcanic rocks of Cenozoic age are particularly abundant in the Central Mountain Region and Basin and Range Province, with less abundant datable plutonic and volcanic rocks of Precambrian age in all three provinces. Sedimentary rocks rarely contain datable minerals and thus absolute ages for them depend on their association with interstratified or intrusive igneous rocks. The "millions of years ago" column in the Geologic Time Scale (Table 4–1) shows the absolute ages of such rocks that have been correlated with relative age units such as periods and epochs on a worldwide basis.

Geomagnetic Time Scale

Another dating and correlating technique has been developed in the last few years which provides a method that may yield even more precise chronologic discrimination of rocks than the relative or isotope dating techniques. The *geomagnetic time scale* is based on the magnetic polarity or orientation of magnetic minerals in rocks. During the molten stage of igneous rocks, magnetic minerals are free to move about freely and tend to become aligned in the melt with their magnetic polar axes parallel to the earth's magnetic field. Upon cooling and crystallization of the magma, the magnetic minerals are "frozen" in position with their north-seeking pole pointing to the earth's north pole. The same orientation process prevails with magnetic particles suspended in water as sedimentary particles, therefore sedimentary rocks also contain a record of the orientation of the earth's magnetic field at the time the rock was formed.

It has been observed that the "frozen" magnetic polarity of rocks, or geomagnetism, varies in rocks of different ages, and commonly exhibits complete reversals of polarity with the north-seeking poles of magnetic particles actually pointing to the earth's present south pole. Since the earth's magnetic field is worldwide in scope, such reversals in the geologic past would have been recorded everywhere in rocks (either igneous or sedimentary) that were forming at any given time. Periods of such magnetic reversals have alternated with periods of normal magnetic field orientation, and the records of the geomagnetic fluctuations are preserved in the rocks. This record shows that the fluctuations were of unequal duration. Certain reversal events are recognizable in different areas due to their duration, as measured by isotopic dating of the containing rocks or by their position in a sequence of alternating reversed or normal magnetic sequences of varying periods of time.

The most complete geomagnetic chronology has been compiled from records in core samples of oceanic sediments and volcanic rocks ranging in age from Recent to Cretaceous, because they represent a continuous record of sedimentation. Rocks exposed on land of equivalent ages show similar sequences of geomagnetic fluctuations therefore can be correlated with a reasonable degree of accuracy with the more complete oceanic records. This phenomenon has permitted the correlation of rocks of terrestrial and oceanic origin which is difficult to do with biostratigraphic techniques because these two environments rarely have the same kinds of organisms present with which correlations could be made. Numerous reversals have been detected in rocks ranging back to the Precambrian and eleven are well documented in rocks ranging in age to 6 million years before present. However, there has been no reversal or change in the magnetic field for the past 700,000 years. Occasionally, during a long normal or reversed phase, the field experienced a reversal for a short period of time. These short periods are called "events," such as the Jaramillo and Olduvai Events in the Matuyama Phase, and are extremely useful as precise age indicators. The isotopic dating of rocks containing records of the geomagnetic phases and events is making possible the development of a worldwide chronology in which the boundaries of the phases can be definitely established. The geomagnetic time scale is considered to be well established for the last 6 million years, but less reliable in older rocks due to uncertain dating and incomplete sequences of sedimentary and igneous rocks. Research in this technique is continuing and will ultimately yield more complete and useful chronology in older rocks.

CHAPTER **5**

SEDIMENTS AND SEDIMENTARY ROCKS

Sedimentary rocks are those which are composed of broken particles of other rocks, of fossil materials that have accumulated in water or air, or of minerals that have formed by chemical precipitation from solution in water. A large portion of Arizona's surface, especially the northern half, is composed of sedimentary rocks such as sandstone, shale or limestone. The walls of the Grand Canyon are carved into several thousand feet of horizontal sedimentary rocks (Figure 5–1), and many of the mountain ranges in southern and western Arizona are made up of tilted layers of sedimentary rocks (Figure 5–2).

Figure 5–1. Horizontal strata in Grand Canyon. View from South Rim toward Bright Angel Canyon and North Rim.

33

Figure 5-2. Tilted Paleozoic marine strata near Miami.

Sedimentary rocks contain important clues to the geologic history of Arizona and often are economically important as host rocks for valuable natural resources such as ground water, oil, gas, coal and even metallic minerals such as copper, gold and uranium. Much of the land in Arizona is covered by sediments such as clay, sand and gravel which have not yet been converted to sedimentary rock. This is especially true in the broad, alluvial valleys between mountain ranges in southern Arizona. In order to understand and fully appreciate the nature and significance of sedimentary rocks we need to consider the processes by which they are formed.

Since sedimentary rocks are composed of sediments, let us consider how sediments are formed. There are three general categories of sediments classified on their mode of origin (Table 5-1). These are clastic, chemical and organic sediments. *Clastic sediments* are broken particles of minerals or rocks that have been removed from rock outcrops by weathering and erosion. Examples of such materials range in size from boulders (greater than 256 mm) to clay size particles (less than 1/256 mm). The size of such particles is dependent on a variety of factors such as original size, hardness, and solubility, but is also directly related to the distance they have been transported by the agents of erosion—water, wind or ice. For this reason the source of sediments can be interpreted for any given sedimentary rock, yielding information on past geography. Coarse materials (gravel) in conglomerates such as the Precambrian Barnes, Scanlan or Hotauta (Figure 5-3) cannot have been transported very far and therefore indicate a hilly or mountainous terrain near the area where the conglomerates were deposited. Sand-sized particles in rocks indicate a more distant source from which sediments may have been removed from igneous, metamorphic or sedimentary rocks. The size range of particles (sorting) and the degree of roundness of particles (Figure 5-4) provide clues to the agents of transportation and the environment of deposition of sediments and sedimentary rocks.

TABLE 5–1. Classification of Sedimentary Rocks

CLASTIC

Particle Size	Sediment	Sedimentary Rock
More than 2 mm	boulder, cobble, pebble	conglomerate
2 mm—1/16 mm	sand	sandstone
1/16—1/256 mm	silt	siltstone
Less than 1/256 mm	clay	shale or mudstone

CHEMICAL

Siliceous	Calcareous	Other
chert	nonfossiliferous limestone	halite
flint	tufa	
chalcedony	travertine	gypsum
jasper		

ORGANIC

Siliceous	Calcareous	Carbonaceous
Radiolarian chert	chalk	coal
Diatomite	fossiliferous limestone	peat
	coquina	

Figure 5–3. Hotauta Conglomerate (younger Precambrian) resting on Vishnu Schist (older Precambrian), North Kaibab Trail, Grand Canyon.

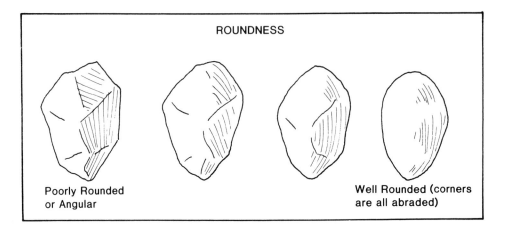

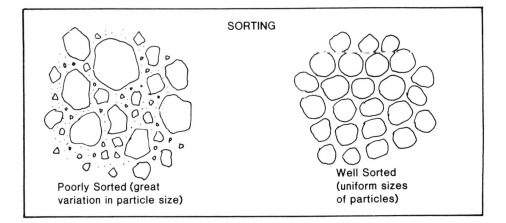

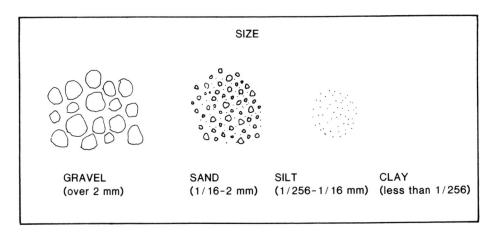

Figure 5–4. Diagram illustrating size, sorting and roundness of sedimentary particles.

As a general rule, the smaller the size and the better the particles are rounded and sorted, the longer the distance they have been transported from their source. Also, sorting and rounding are generally improved in wind transported materials and by wave action in marine environments as compared to stream deposited sediments.

For example, the coarse particles of sand in the Cambrian age Tapeats Sandstone in the Grand Canyon are poorly sorted and not well rounded, suggesting that they were transported only a short distance before being deposited. This can be substantiated by the observation that the Tapeats rests on an erosional surface of Precambrian igneous and metamorphic rock from which the sand grains were derived (Figure 5–5).

In contrast the medium-grained, well-sorted sand of the Cretaceous age Toreva Sandstone of Black Mesa, which extends eastward and northward for hundreds of miles, rests upon the Mancos Shale which contains essentially no sand. This sand must have been transported hundreds of miles from its source and then deposited on the black mud bottom of the Mancos sea. The distribution of this vast volume of sand began at an outcrop of granite, or other quartz-bearing rock, where particles were weathered loose then moved by gravity and running water along streams to the ocean, where wave and current action spread it over a large area of shallow sea floor. These processes of weathering and erosion slowly abraded the particles to small, rounded sand grains and sorted them to uniform sizes by the constant winnowing action.

Figure 5-5. Tapeats Sandstone (Cambrian) resting on erosion surface on Precambrian granite, near Payson, at crossing of Hwy 87 and East Verde River.

Stream deposits are relatively coarse grained near their source. For example, the Triassic age Shinarump Conglomerate exposed near Cameron was derived from the weathering and erosion of mountains called the Mogollon Highlands in central Arizona, just a few tens of miles away (Figure 5–6). Further from their source the sediments are finer grained, such as the sandstones of the Permian age Esplanade Sandstone in the Grand Canyon and the Schnebly Hill Formation in Oak Creek Canyon which was transported hundreds of miles from the nearest possible exposed source area (Figure 5–7).

The best sorting of sand occurs under conditions of wind deposition because the carrying capacity of wind is much less than of water. For a given wind velocity all particles of a certain maximum size plus all smaller particles will be picked up and moved along. With a slight decrease in wind velocity the coarser fraction can no longer be suspended, and therefore drops out as sediments, while the finer fractions are winnowed out and carried away for deposition elsewhere. The result is fine-grained, well-sorted, sands, such as those in the Permian Coconino Sandstone of the Grand Canyon (Figure 5–8), the DeChelly Sandstone of Canyon DeChelly and Monument Valley, and the Jurassic Navajo Sandstone at Lake Powell.

Figure 5–6. Shinarump Conglomerate (Triassic) near Cameron. Note abundant large, well-rounded particles (2 inches or more) and poor sorting, typical of sediments that have been transported by strong water currents from a nearby source.

Figure 5-7. Schnebly Hill Formation near Sedona. It consists of fine to medium grained, well-sorted sand, typical of sediments that have been transported a long distance.

Figure 5-8. Coconino Sandstone (Permian) cross-bedding resulting from wind transport of sand grains. Photo in Oak Creek Canyon.

The rocks formed from fine-grained (1/256–1/16 mm) sediments are called shales if they occur in thin layers or laminae, or mudstones if they are not thinly laminated. Such deposits are usually formed in aquatic environments, either marine or nonmarine. Examples of extensive deposits in Arizona are the Precambrian Hakatai, Dox and Pioneer Shales; Cambrian Bright Angel Shale; Devonian Morenci Shale; several Pennsylvanian formations; Permian Hermit Shale; Triassic Moenkopi Formation; Cretaceous Mancos Shale; and several shale and mudstone rock units of Cenozoic age such as the Verde Formation. As a rule of thumb, the dark gray to black shales of Arizona are of marine origin and the red, yellow or brown shales are of nonmarine or transitional origin. The contained fossils provide definitive evidence of environment of deposition. For example, the Triassic red shales generally contain terrestrial organisms such as land plants and reptiles, and the Cretaceous Mancos Shale contains oysters, cephalopod mollusks, sea urchins, sharks and corals indicative of marine environments. Dark gray to black shales of Paleozoic age generally contain brachiopods, crinoids, bryozoans and trilobites, all of which indicate marine environments. Such fine-grained sediments generally suggest deposition in marine environments some distance from shorelines beyond which there was inadequate current and wave action to transport sand-size or coarser particles.

Another way in which sediments are formed is through biological activity. A visit to any seashore beach where the surface is littered with whole or broken shells would provide a most obvious example. Such shell deposits may be several feet thick and extend far out to sea forming a sedimentary deposit that will ultimately become a limestone. Other organic limestone deposits may consist of masses of reef-forming corals or other framework-building organisms, the skeletons of tiny single-celled protozoans called foraminifera, or finely laminated layers of limestone called stromatolites which are precipitated from water by the photosynthetic activity of blue-green algae.

Examples of Arizona limestones that were formed primarily by these processes are the Devonian Martin Formation with abundant corals, brachiopods, mollusks and algal stromatolites (Figure 5–9); the Mississippian Redwall and Escabrosa Limestones with corals, crinoids, trilobites, brachiopods, bryozoans, mollusks and foraminifera (Figure 5–10); the Naco and other Pennsylvanian formations with abundant crinoids, brachiopods, mollusks, corals and foraminifera (Figure 5–11); the Cretaceous Mancos Shale with ammonites, corals and oysters (Figure 5–12); and the Pliocene Verde Formation whose limestones are primarily due to the photosynthetic activity of plants (Figure 5–13).

Coal, peat and lignite beds are rocks formed by the accumulation and burial of large amounts of plant tissue. The probable environments of most of the coal deposits are in ancient swamps similar to the Okefenokee Swamp in Georgia where dead trees fall into water rich in humic acid and are soon buried by mud, both of which inhibit decay or destruction by bacteria thereby preserving the carbon of the wood. In Arizona we have a small amount of coal in Pennsylvanian rocks along the Mogollon Rim, and extensive deposits in Cretaceous rocks of Black Mesa (Figure 5–14).

Other sedimentary rocks of biological origin include some chert formed by the accumulation of siliceous (SiO_2) skeletons of single-celled protozoans called radiolarians, sponge skeletons or single-celled algae called diatoms. No radiolarian cherts have been identified in Arizona; however, sponges are common in the Kaibab Limestone which commonly form the nucleus of siliceous concretions (Figure 5–15). Diatomites of Tertiary age are found in the nonmarine deposits of the Safford Basin near Duncan and in the Verde Basin near Camp Verde.

Figure 5-9. Fossil corals and brachiopods in the Martin Formation, of Devonian age, southwest of Camp Verde.

Figure 5-10. Fossil brachiopods, crinoid and trilobite in the Redwall Limestone of Mississippian age, Mogollon Rim.

Figure 5-11. Fossil brachiopods in the Naco Formation of Pennsylvanian age, near Payson.

Figure 5-12. Fossil ammonites in the Mancos Shale of Cretaceous age, near Kayenta.

Figure 5-13. Fossil plants in the Verde Formation of Miocene and Pliocene age, near Clarkdale.

Figure 5-14. Coal beds in Dakota Sandstone of Cretaceous age, Coalmine Canyon near Tuba City. Vertical exposure is 4 meters (12 feet) high. Petrified logs and leaf impressions are common in the sandstone above the coal.

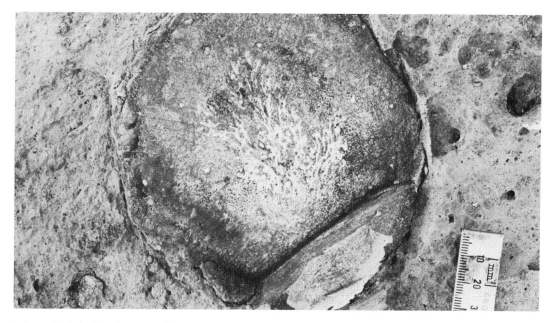

Figure 5–15. Fossil sponge in siliceous concretion within the Kaibab Limestone, Hermit Trail, Grand Canyon.

Chemical sediments may include limestone ($CaCO_3$), chert (SiO_2), gypsum ($CaSO_4$) and halite ($NaCl$) which form when chemical or physical changes cause salts dissolved in water to precipitate. This precipitation may occur in standing bodies of water (primary mineralization), or it may occur in ground water occupying space within existing rocks or sediments (secondary mineralization). Examples of primary mineralization in Arizona include dolomite and chert within the Devonian Martin Formation, gypsum and halite in the Permian Toroweap, the Permian Supai Group in the Holbrook Basin, and the Cenozoic basins near Picacho, Luke Field, Red Lake and Verde Valley. Deep drilling in several Cenozoic basins in Arizona has discovered thicknesses of up to 10,000 feet of gypsum and halite.

Sedimentary Structures

Additional clues to the environments of deposition of sediments and sedimentary rocks may be found in *sedimentary structures* that were formed during deposition of the sediments.

Sedimentary structures are features such as bedding, ripple marks or mud cracks which form in sediments at or near the time they were deposited, while they are soft and unconsolidated. Most known types of sedimentary structures have been observed in modern sediments, therefore the specific environment of their formation can be observed. Based upon this knowledge of the environmental significance of sedimentary structures, we can apply the Principle of Uniformitarianism and interpret the depositional environment of the sediments that compose sedimentary rocks. Sedimentary structures are formed by inorganic and/or organic processes. Examples of inorganic sedimentary structures include ripple marks generated by waves and currents (Figure 5–16), cross-bedding (Figure 5–8) and mud cracks (Figure 5–17). Inorganic sedimentary structures may

Figure 5-16. Ripple marks in Shinarump Conglomerate north of Cameron.

Figure 5-17. Mud crack casts in Tertiary age sedimentary rocks on the north side of Tempe Butte.

be encountered in sedimentary rocks of any age because they are formed by physical processes such as gravity, water currents, waves, wind, wetting and drying which have not varied significantly throughout earth history. For example, mud cracks and ripple marks may be observed forming today and they are common in sedimentary rocks as old as Precambrian such as the one-billion-year-old Hakatai Shale in the Grand Canyon.

Organic sedimentary structures are those formed by the activities of living organisms such as worm burrows (Figure 5–18), dinosaur tracks (Figure 5–19), elephant tracks (Figure 5–20), plant root molds and many others. Obviously such organic sedimentary structures as dinosaur tracks would only have been formed during the time that dinosaurs lived on the earth—the Mesozoic Era. Likewise, elephant tracks were formed during the late Cenozoic after which time the elephants are known as fossils. Conversely the burrows of soft-bodied worms are commonly found in sedimentary rocks as old as Cambrian. Organic sedimentary structures are generally used to interpret depositional environments and the activities of organisms rather than for age determinations as are the actual remains or body fossils of organisms. Table 5–2 lists the most common sedimentary structures and the environmental conditions indicated by each.

Figure 5–18. Burrows dug by unknown bottom-dwelling organisms in Kaibab Limestone (Permian), Hermit Trail, Grand Canyon.

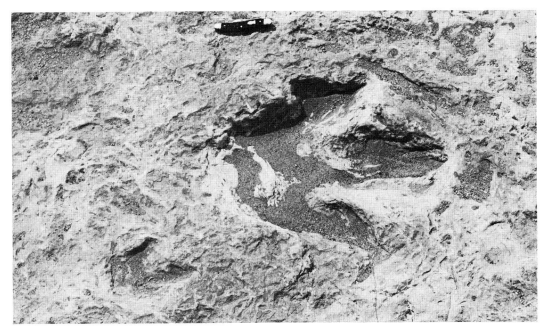

Figure 5-19. Dinosaur tracks in Kayenta Formation (Triassic) near Tuba City, Arizona. Large track is imprint of hind foot, small track is imprint of front foot of bipedal carnivorous dinosaur. Large track approximately 30 centimeters (one foot) long.

Figure 5-20. Mastodon (elephant) tracks in Verde Formation (Miocene) limestone near Camp Verde. Tracks are about 45 centimeters (18 inches) in diameter.

TABLE 5-2

Sedimentary Structure	Description and Environmental Significance
1. Horizontal bedding (Figure 5-21a)	Sedimentation in standing bodies of water or air.
2. Cross bedding (Figure 5-21b)	Sedimentation in water or air currents. Dip direction of cross beds indicates current direction.
3. Graded bedding (Figure 5-21c)	Gradual change of sediment size within a layer from coarse at bottom or fine at top. Indicates deposition in turbidity flow.
4. Imbrication (Figure 5-21d)	Uniform direction and angle of inclination of disk shaped pebbles or cobbles within a layer of sediments. Indicates current direction.
5. Concretions (Figure 5-21e)	Irregular or subspherical shaped, more resistant masses within sedimentary rocks.
6. Geodes (Figure 5-21f)	Hollow, globular bodies with well developed internal crystals radiating from wall toward center.
7. Load casts (Figure 5-21g)	Irregularities of the under side of sand or gravel beds caused by its sinking into soft mud upon which it was deposited.
8. Flute casts (Figure 5-21h)	Sediment filling of elongate grooves scoured from bedding plane surface by current action. Indicates current direction.
9. Mud cracks (Figure 5-21i)	Polygonal cracks in sediments or sedimentary rocks due to alternating wet and dry conditions in ponds, lakes, or tidal flats.
10. Ripple marks (Figure 5-21j)	Elongate, alternating ridges and troughs in sediment caused by waves or currents in water or air.
11. Rain or hail prints (Figure 5-21k)	Circular pits in bedding plane surface caused by impact of rain drops or hail. Indicates subaerial exposure.
12. Biogenic sedimentary structures (Figure 5-21l)	Footprints, trackways, root molds, burrows or fecal castings of organisms in sediment or on bedding planes. Provides evidence of activity and environment of deposition.

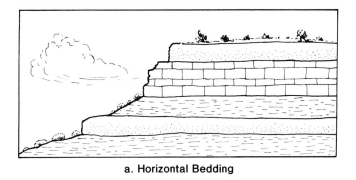

a. Horizontal Bedding

b. Cross-Bedding

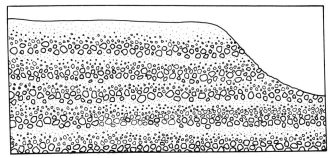

c. Graded Bedding

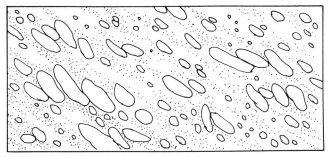

d. Imbrication

Figure 5–21. Drawings of sedimentary structures discussed in Table 5-2.

Figure 5–21—*Continued.*

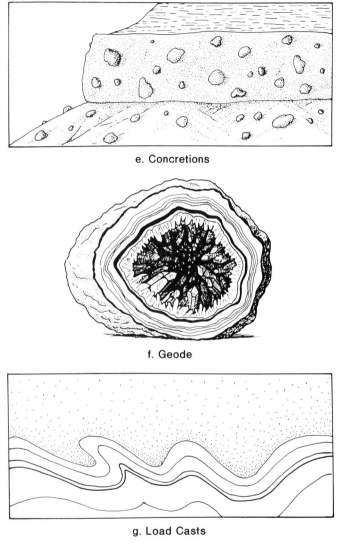

e. Concretions

f. Geode

g. Load Casts

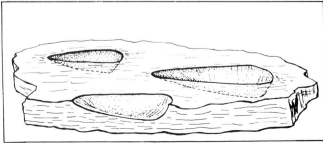

h. Flute Casts

Figure 5–21—_Continued._

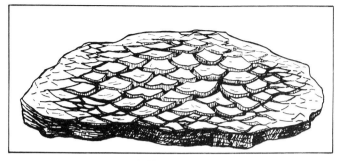

i. Mudcracks

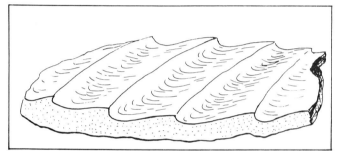

j. Ripple Marks

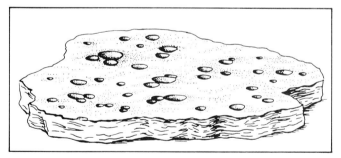

k. Rain prints

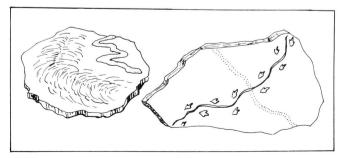

l. Biogenic Structures

Sedimentary Environments

The interpretation of ancient sedimentary environments is essential to an understanding of the geologic history of Arizona. Each layer of sedimentary rock contains many clues to the particular environment in which it was deposited. Such clues include the mineral and chemical composition of the sediments; the texture, including grain size, sorting, and roundness; and sedimentary structures. The techniques of interpreting sedimentary environments are based on the fundamental Principle of Uniformitarianism—comparison of such features with those that can be observed forming in modern sedimentary environments.

Sedimentary rock units in Arizona provide evidence of changing conditions during the past two billion years of earth history. In order to interpret those past conditions let us consider the following modern sedimentary environments to be used as models for comparison.

Sediments that are being transported by running water or wind may be deposited in any place where those agents of transportation lose the dynamic energy that is necessary to keep the sediment load in motion. Many examples of sedimentary environments exist, but they can be classified in three general categories for the purposes of our discussion. *Continental environments* are those areas of land located beyond the direct influence of ocean water. They include terrestrial (or dry land) and aquatic (lakes, streams, springs) environments (Figure 5–22a). *Transitional environments* are those areas that are periodically covered by marine water, such as beaches and tidal flats or aquatic areas such as bays and estuaries where there is a mixing of marine and fresh water (Figure 5–22b). *Marine environments* include all areas of the earth's surface that are constantly submerged beneath ocean water. These environments are further subdivided based upon water depth (Figure 5–22c). Marine environments are also classified into life zones to which living or-

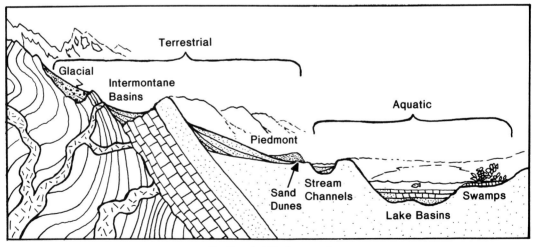

a. Continental Environments

Figure 5–22. Depositional environments. Diagrams of a. continental, b. transitional and c. marine environments.

Figure 5-22—*Continued.*

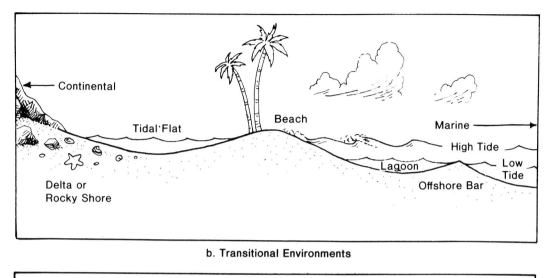

b. Transitional Environments

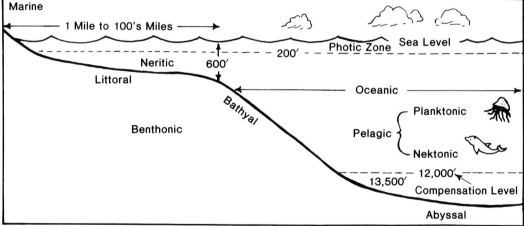

c. Marine Environments

ganisms become adapted. For example, the ocean floor is called benthonic, the water above the ocean floor is called pelagic, and subdivisions of each are shown in Figure 5-22c.

In addition to the physical characteristics of sedimentary rocks, their fossil content also provides evidence of the environment in which the sediments were deposited. There are obvious differences in the kinds of organisms living on land and in the ocean. The plants and animals that live in lakes and streams are different from those that live in the oceans, even though both are aquatic environments. All living organisms are adapted to specific conditions (i.e., temperature, water needs, water salinity, food supply, water depth, type of substrate) and are unable to survive or prosper under other conditions. For this reason, once we know the ecologic limitations of a plant or animal, we can use that knowledge to interpret the ecologic conditions that prevailed in the past (paleoecology) if it occurs as a fossil. For example, the fossils found in the Mississippian Redwall or Escabrosa Limestone include corals, brachiopods, crinoids and cephalopod mollusks, all of which are strictly marine animals today (Figure 5–10). In contrast, the Pliocene Verde Formation contains horses and mastodons, which were terrestrial mammals, and snails and clams of the same species that live in fresh water streams and lakes today (5–13).

The Cretaceous age Dakota, Toreva and Wepo formations in Black Mesa contain coal beds (Figure 5–14) that were formed from the accumulation of thick layers of tree trunks, leaves and other vegetation that must have grown in swampy coastal plain environments where they could be preserved by burial in mud. We know they were coastal plain environments because they are closely associated with marine deposits containing oysters, ammonites, corals and other marine organisms.

The occurrence of layers of sediment in a vertical sequence, representing different environments of deposition can be used to interpret changing conditions through time in that area. For example, the Cretaceous strata at Coalmine Canyon near Tuba City, and elsewhere in Black Mesa, grade vertically from stream channel sandstones to carbonaceous shales to coal beds (all terrestrial environments) which are overlain by oyster "reefs" (a near-shore brackish marine environment) and then black shales and sandstones of the Mancos Shale with normal marine organisms such as snails, bivalves and ammonites which must have lived some distance offshore (see Figure 5–12). These changing environments clearly depict a marine transgression over the area during the Cretaceous Period (Figure 5–23).

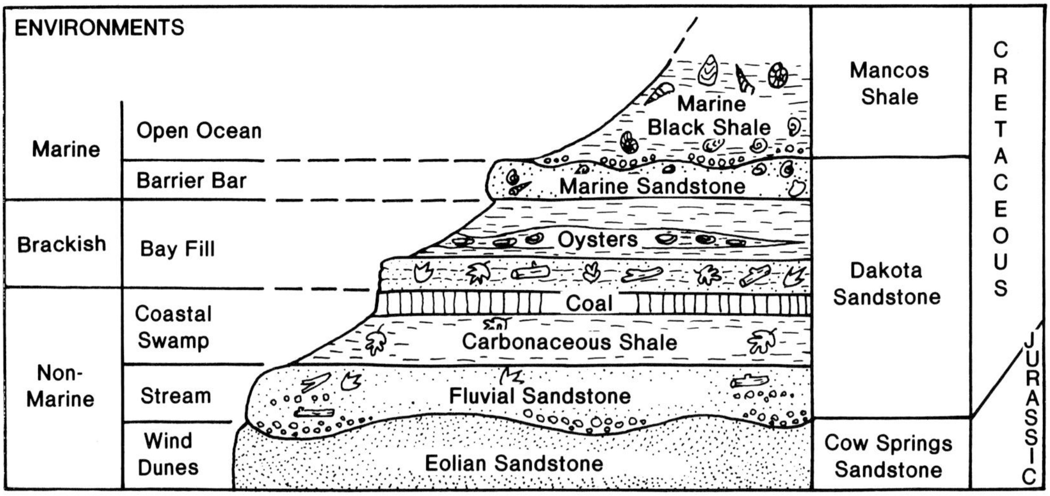

Figure 5–23. Marine transgression in the Black Mesa area during late Cretaceous time, depicted by upward change in lithology and fossil content from the nonmarine Dakota Sandstone to the marine Mancos Shale.

Another example of changing environmental conditions can be observed along the Mogollon Rim in the vertical transition from the Pennsylvanian Naco Formation with a diverse fauna of marine organisms (bryozoans, brachiopods, crinoids, etc.) into the Permian Supai Group containing predominantly redbed sediments and fossils of land plants (ferns, scale trees, horsetails). This sequence clearly depicts a marine regression in central Arizona from Pennsylvanian to Permian time. Other examples of changing environments are described in Chapters 10 and 11.

TABLE 5–3.   Fossils as Indicators of Environment

Fossil Group	Ecologic Significance
Land Plants	Those having well-differentiated roots, stems and leaves indicate continental environment.
Microscopic plants	
Diatoms	Fresh water or marine, different genera in each.
Calcareous algae	Mostly marine, shallow, well-lighted, clear water, may be supratidal—transitional environment, some in fresh water lakes.
Foraminifera	Marine, larger forms generally indicate shallow, warm water environment, but many types live in deeper, colder water.
Sponges	Marine, probably shallow, clear water.
Corals	Marine, most indicate shallow, warm, clear, normal salinity (3.5%) dissolved salts.
Bryozoans	Marine, shallow, well-oxygenated water.
Brachiopods	Marine, probably clear water and a firm, stable bottom. Most diverse and abundant in Paleozoic age rocks.
Bivalvia	Marine or fresh water, different genera in each. Marine forms much more varied than fresh water. Most diverse and abundant in Cenozoic age rocks.
Gastropods	Marine or fresh water, terrestrial. Predominantly marine, most diverse and abundant in Cenozoic age rocks.
Cephalopods	Marine, mostly floating or swimming forms. Most abundant and diverse in Mesozoic age rocks.
Scaphopods	Marine, shallow to deep water but offshore, soft bottom.
Trilobites	Marine, generally quiet water and soft, muddy bottom conditions. Exclusively in Paleozoic age rocks.
Ostracods	Marine, brackish or fresh water. Most abundant and diverse in marine water but commonly found in fresh water lake deposits.
Insects	Terrestrial, preserved only in most favorable conditions of sedimentation.
Crinoids	Marine, shallow or deep well-aerated water, most abundant in Late Paleozoic.

TABLE 5–3. *Continued.*

Fossil Group	Ecologic Significance
Echinoids	Marine, most indicate shallow quiet water conditions.
Fish	Marine or fresh water, great variety of aquatic environments.
Amphibians	Aquatic fresh water, fossils rare.
Reptiles	Many land environments, some aquatic and marine. Dinosaurs and marine reptiles abundant in Mesozoic.
Mammals	Mostly land, diet and habits can be determined by types of teeth and limbs. Small types appeared in Mesozoic; abundant and diverse in Cenozoic.

CHAPTER 6

IGNEOUS AND METAMORPHIC ROCKS

Igneous Rocks

To understand igneous rocks perhaps it is best first to imagine what *magma* is like. Being melted rock, it is very hot, greater than 750°C (1382°F) and often more than 1000°C (1832°F). It is a liquid, but a very sticky one (this property of stickiness, called *viscosity* varies according to the chemical composition of the magma), that originates by the melting of rocks deep in the crust or in the upper mantle (see Chapter 7). Since it is slightly less dense than the rocks from which it is melted, it tends to rise slowly by displacing the rocks above it. Often it cools and crystallizes before reaching the surface, but if it rises all the way it erupts as lava producing a volcano, an event repeated many times in the past across Arizona (Figure 6–1; also see Chapter 13).

The eruption of hot lavas is direct evidence that temperatures are much greater at depth than at the surface. Measurements in deep mines and boreholes indicate that temperatures in the crust rise at between 2 and 3°C for every 100 meters (330 feet) depth of rock. It would seem logical then to assume then that anywhere on earth, if you went deep enough, the temperature would be high enough to have produced magma. However, this is not the case. Rather, magma seems to be generated only at certain points and narrow zones throughout the earth. This apparent paradox is explained by the fact that as depth increases so does the pressure from overlying rock, and that increasing the pressure on a rock also tends to increase its melting point. Thus, increased pressure counteracts the increased temperature and usually predominates, so that melting does not occur. However, in certain active regions of the earth (see Chapter 7) heat builds to such a degree that the rock does melt.

To visualize the interplay between temperature and pressure on a rock, picture the individual atoms in a single crystal of some mineral, neatly arranged in rows in the crystal lattice, vibrating quietly. As heat is added to the crystal the atoms become more energized and vibrate more strongly. Eventually the vibrations are so great that the crystal does not have the strength to hold together its individual atoms and they begin flying apart. This dissociation of the crystal is melting. Bringing pressure into the picture tends to dampen the vibrations caused by increased temperature and to hold the atoms in their place, thus inhibiting melting.

As melting begins, large portions of the rock are still crystalline, but as the process continues the remainder of the solid material is either melted or the magma is drawn away from it. The magma coalesces as a liquid mass surrounded by unmelted country rock, known as a *magma chamber*.

When the magma in the magma chamber reaches some critical volume it will begin to rise up through the denser rocks above it, or intrude them (Figure 6–1). Intrusion of the magma is slow and difficult, occurring by one of several related processes. One is melting and assimilation

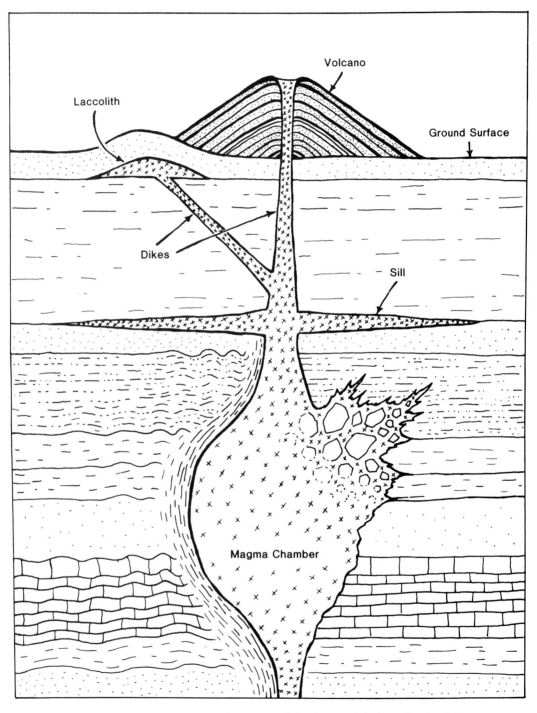

Figure 6–1. Classification of igneous rock bodies, and their mode of origin.

of the overlying rock by the rising magma body. This may be thought of as the intrusion eating its way upwards. Sometimes magma will squeeze into cracks in the roof of the magma chamber dislodging large chunks of the rock which sink into the magma and eventually are themselves melted and consumed. This process is known as *stoping,* a term first used for a type of underground mining in which the roof of a tunnel is blasted loose and dropped down so it can be hauled away by the miners (see Chapter 15). Another method of intrusion is by *displacement,* that is, by magma pushing the overlying rocks on ahead of it or off to the sides. This is aided by the softening of the walls of the magma chamber by heat from the magma, so that the surrounding rocks behave plastically and are more easily pushed aside by the advancing magma.

As magma intrudes into cooler regions of the crust it loses heat and eventually begins to crystallize, passing through a phase in which there is a magma-crystal mixture before becoming completely solid. If the magma does not reach the surface of the earth before it completely crystallizes, the resultant body of rock is called a *pluton.* Pluton is the general term for any igneous body that congeals at depth, regardless of its size or shape. If an igneous body is greater than 100 km² (36 mi²) in areal extent, it is then called a *batholith;* it may be a single pluton, but more often is a composite of a number of individual, smaller plutons.

Sheet-like or tabular plutons are given different names. Those that are intruded between layers of rock, usually between bedding in sedimentary rock are called *sills* (Figure 6–1). Sills are particularly well displayed in the walls of the Salt River Canyon where igneous rocks have intruded the beds of the Apache Group (Figure 6–2). *Dikes* are those tabular plutons that intrude across

Figure 6-2. Cliff forming basaltic sills in Mescal Limestone of younger Precambrian age, Salt River Canyon.

Figure 6–3. Basalt dike (one billion years old) instrusive into Hakatai Shale of younger Precambrian age, Hance Rapids, Grand Canyon.

layers of rock, usually along fractures (Figure 6–3). Finally, a rather unusual pluton, called a *laccolith,* forms when magma intrudes between layers of rock at a shallow depth causing an upward bulge in the layers above, while the floor of the pluton remains flat (Figure 6–1). Navajo Mountain on the Arizona-Utah border is probably an example of a laccolith.

If magma rises to the ground surface before crystallizing it will pour out or *extrude* as a volcanic eruption. Eruptions are of two basic types; *lava eruptions,* in which the magma pours out across the terrain as a sticky liquid, and *pyroclastic eruptions,* which are explosive so that the magma is blown into many small fragments. The first type is familiar to most people from the series of eruptions in recent times on the big island of Hawaii, whereas the second type is exemplified by Mt. St. Helens, Washington, in 1980. Volcanic eruptions and their resulting deposits are highly varied in detail. Arizona has suffered widespread volcanism in the fairly recent geological past, which has contributed much to the present scenery of the state. The fascinating subject of volcanic processes will be taken up in detail in Chapter 13, but now we will consider the general classification of igneous rocks.

Classification of Igneous Rocks

The goal of any classification system is to divide all the various individuals in some group into categories along lines of related characteristics. Igneous rocks are subdivided according to two basic parameters: 1) the size of mineral grains and 2) the chemistry or mineralogy of the rock. A simple classification is given in Table 6–1.

TABLE 6–1. Classification of Igneous Rocks According to Chemical Composition and Grain Size

	Silicic		Intermediate	Mafic	Ultramafic
Coarse-grained	Granite	Granodiorite	Diorite	Gabbro	Peridotite
Fine-grained	Rhyolite	Dacite		Andesite	Basalt

Mineral Grain Size. The size of crystals depends on how quickly the magma cools and crystallizes. Slowly cooling magma produces large crystals and quickly cooling magma, small crystals. The distinction between coarse-grained (or phaneritic) rocks and fine-grained (or aphanitic) igneous rocks is whether the individual crystals can be seen with the unaided eye. Magma will cool much more quickly in the air or on the earth's surface than it would have in the subsurface. In general then, the fine-grained rocks in the classification are extrusive and the coarse-grained rocks are intrusive. There are exceptions however, for if a magma intrudes into shallow, cold rock it may cool quickly enough to be aphanitic. If a magma is cooled very quickly, particularly a rhyolite magma, it may not have a chance to crystallize at all and instead will quench as a natural glass, called *obsidian.*

In a fashion analogous to a rock that is only partially melted on its way to becoming magma, as a magma begins to crystallize at depth there will be relatively few crystals growing in it. If cooling is slow the crystals may reach a length of several millimeters or even centimeters (2.5 centimeters = 1 inch). Then if the magma intrudes upward into cooler rocks or extrudes it will cool more quickly, and the resulting crystals will be smaller. The resulting igneous rock, called a *porphyry* (see Figure 2–1c), will have some large crystals surrounded by distinctly smaller ones, be they large enough to see with the unaided eye or not. The larger crystals in a porphyry are known as *phenocrysts.*

Chemistry. Remember from an earlier discussion that there are only eight elements in the earth's crust that account for most of its volume. These elements are oxygen (O), silicon (Si), aluminum (Al), potassium (K), sodium (Na), calcium (Ca), iron (Fe) and magnesium (Mg). All magmas are also composed primarily of these eight elements in roughly the same proportions as in the crust. However, the abundances of the elements relative to one another does vary significantly from one magma to another. The elements silicon, aluminum, potassium and sodium tend to vary as a group, as do calcium, iron and magnesium. When the former group is relatively enriched the magma is said to be *silicic.* When the latter is relatively enriched the magma is *mafic,* and when magma chemistries are in between these two extremes they are *intermediate.*

The chemistry of a magma directly affects the type of minerals that will crystallize from it. At the mafic end of the spectrum the minerals are likely to be olivine and pyroxene, minerals rich in iron and magnesium, and a form of plagioclase feldspar rich in calcium; the resultant rock is dark colored. At the silicic end, the rock is likely to contain orthoclase, the potassium-bearing feldspar, a sodium-rich form of plagioclase, and quartz; the resultant rock is light colored. With coarse-grained igneous rocks, identification of the minerals and their relative abundances serves to identify the rock (Table 6–2). With volcanic rocks the mineral grains are too small to distinguish. Reasonable identification of igneous rocks may be made in the field, without chemical analyses, by the color of the rock on a freshly broken surface because color is a rough guide to the rock's chemistry. Rhyolites and granites tend to be light colors, tans, grays, pinks, whereas basalts and

TABLE 6–2. Classification of Coarse-grained Igneous Rocks According to Minerals Present

Granite	Granodiorite	Diorite	Gabbro
Quartz	Quartz	Plagioclase (very minor or no quartz and K-feldspar)	Plagioclase
K-feldspar > plagioclase	Plagioclase > K-feldspar		Pyroxene
Biotite	Biotite or hornblende	Hornblende Biotite common	Olivine common

diabases are often dark gray. Andesites are darker than rhyolites and lighter than basalts. In practice this system is imprecise for some rhyolites, particularly the more glassy varieties, are black, and basalts, especially if they are weathered somewhat, may be medium gray in coloration, similar to andesite. Phenocryst composition may sometimes be a clue with quartz usually indicating a rhyolite, and olivine indicating a basalt.

Experience is the best teacher, look at pictures, look at museum collections, but most of all look at the rocks in the field.

Metamorphic Rocks

Metamorphic rocks have been formed during several episodes in the geological history of Arizona. The oldest rocks in the state are of this type, cropping out in the bottom of the Grand Canyon and in numerous locations in the mountainous region south of the Colorado Plateau. Metamorphic rocks are derived from pre-existing rocks, by mineral changes in the solid state, when they are subjected to increased temperature and pressure. Either sedimentary or igneous rocks may be metamorphosed, but the latter generally show little effect since they are formed at high temperatures to begin with.

The minerals in a sedimentary rock in general are stable at temperatures and pressures found at the surface of the earth. As these rocks are buried, pressure will increase due to the overlying rocks, and temperature will increase due to the greater depth. As this happens the minerals become unstable under the new conditions. That is, the crystal lattices begin to break down and regroup, or *recrystallize,* as new minerals. Following the example of melting, imagine the atoms in a crystal lattice, neatly arranged, quietly vibrating. As temperature is increased the vibrations become more energetic until eventually the atoms begin to disassociate. Unlike melting, however, as the atoms come apart they immediately recombine to form a new mineral that is more stable at the new temperature and pressure conditions.

As heat and pressure are increased a rock will proceed through a series of metamorphic reactions. The minerals that form at each step are characteristic of a particular temperature and pressure, so that proper identification of the minerals will indicate the maximum temperature and pressure reached by the metamorphic rock. If enough heat is applied, eventually the rock will melt, magma will be produced, and an igneous rock will form when the magma again crystallizes.

Minerals that have formed under high temperature and pressure deep within the earth, do not reverse the process and change to low temperature and pressure minerals as they are uplifted near or to the surface. As temperatures are raised, it is the added energy that drives the metamorphic reactions. When temperatures are lowered, that energy is removed from the system and the high-temperature minerals stay frozen as they are. Another reason that back-reactions do not occur is that during metamorphism water is expelled from the minerals with which it had been combined, and escapes to the surface. Once the water is lost, it is not available to recombine at lower temperatures.

Metamorphic rocks are of two basic types, contact and regional. *Contact metamorphism* occurs in fairly narrow zones around intrusions, where the intruded rock is baked by the heat of the adjacent magma. *Regional metamorphism,* as the name implies, occurs throughout large volumes of rock. Under certain conditions (see Chapter 7), large sedimentary basins subside as they are filled with sediments. When the sedimentary rocks at the bottom of the pile reach great enough depth metamorphic reactions commence. Unlike contact metamorphism, which is primarily a heating phenomenon, regional metamorphism involves both heat and pressure.

The classification of metamorphic rocks is not as straightforward as with igneous or sedimentary rocks. In part it depends on what the original rock was, and in part on what form or *texture* the mineral grains take in the rock.

If the original rock is composed of only one type of mineral, then the resulting metamorphic rock is mineralogically simple. When a limestone is metamorphosed, it becomes *marble.* A clean quartz sandstone becomes *quartzite.*

Shales, on the other hand, are composed of a variety of clays, minerals that are rich in combined water. As metamorphism progresses, micas grow from the clays. Mica is a platy mineral, and all the individual plates tend to grow with the same orientation. At the onset of growth the micas are too small to be seen, but their parallel orientation permits the rock to break or *cleave* in thin sheets. This type of rock is *slate,* familiar as roofing material or, in days past, as blackboards. Both materials are made possible by the property of cleavage inherent in the *slate.* When the micas grow large enough to be seen, the rock is called a *schist* (Figure 6–4). Schists also usually

Figure 6-4. Highly folded schist and gneiss in older Precambrian rocks in the Sierra Estrella Mountains.

contain quartz and feldspar, but the aligned mica flakes dispersed among the other minerals are what give the rock its character. When metamorphism reaches its highest levels the minerals tend to segregate into discrete bands of quartz, feldspar and mica or hornblende (Figure 6–5). This type of rock is called *gneiss* (pronounced "nice").

In other cases if the original rock is known, its metamorphic equivalent is designated by adding the prefix "meta," for example, metavolcanic rock or metagabbro. When a basalt is metamorphosed the greenish minerals chlorite and epidote are formed and the rock is sometimes called *greenstone*. At higher metamorphic grades when the most abundant mineral is amphibole the rock is called an *amphibolite*.

Figure 6–5. Dike of Zoroaster Granite intrusive into Vishnu Schist (both older Precambrian), Phantom Ranch vicinity, Grand Canyon. Much of the rock in this photo would be classified as gneiss.

CHAPTER 7

TECTONICS

Thus far we have looked at some of the rocks of Arizona that are exposed at the ground surface as outcrops, and discussed general classifications and the processes responsible for their creation. The next step is to place Arizona in a regional and global context, for the state, with all its geological wonders, is no more than a postage stamp on a mountain range that extends from Alaska to the tip of South America. The processes that built such an imposing construct are truly global.

We will begin on the broad scale, with the main regions of the earth's interior, then discuss the processes that are occurring in the outer regions and that are responsible for such features as the ocean basins and the major mountain chains. Finally we will touch on the evolution of western North America to show where Arizona fits in the larger picture.

The Earth's Interior

The earth contains three major layers, the core, the mantle and the crust (see Figure 7–1). Because most rocks at the surface were formed in the outer few tens of miles of the earth, and the deepest rocks which have made their way to the surface have come from only a few hundred kilometers down, what we know of the thousands of kilometers below that depth is inferred from signals that emanate from or pass through that region. When good-sized earthquakes occur they send out waves which travel throughout the globe. These waves are affected differently according to whether they have passed through only the crust; through the crust and mantle; or the crust, mantle and core.

The people who study these earthquake or *seismic waves* are called seismologists. They have defined the boundaries between the three layers, because marked changes in seismic wave velocities are observed at these levels. The *core,* which begins at depths of about 2900 km (1750 miles), is thought to be composed of iron and nickel, similar to metallic meteorites. The outer portion is thought to be liquid and the inner portion to be solid. The *mantle,* which encloses the core, is composed primarily of oxygen, silicon, iron and magnesium. In the igneous rock classification this material is ultramafic, and at least the outer portion of the mantle is probably mostly olivine, an iron-magnesium silicate mineral.

The third layer, a thin skin covering the earth, is called the *crust*. It is composed primarily of the eight elements discussed in the first chapter, but these are quite unevenly distributed. There are two fundamentally different types of crust, that which underlies the ocean basins and that which makes up the continents. *Oceanic crust* is consistently about 8 km (5 miles) thick and of

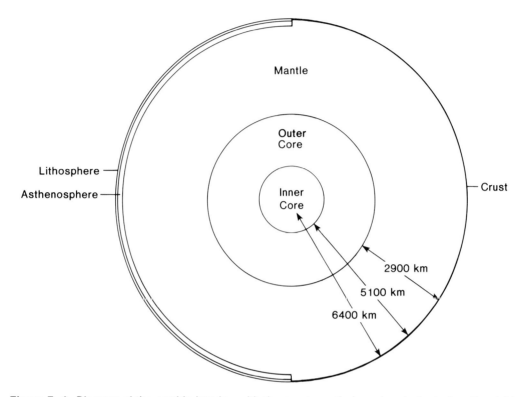

Figure 7–1. Diagram of the earth's interior, with the crust-mantle boundary indicated on the right half and the asthenosphere and lithosphere indicated on the left.

mafic composition, that is, either basalt or gabbro. *Continental crust* varies in thickness from about 20–60 km (12–36 miles), averaging 35 km (21 miles). The rocks of the continents are also highly varied as we have seen from the first several chapters. Since crustal rocks have a lighter density than the mantle, in a sense they float on it. The continents float higher than the ocean basins because they are not as dense, but also because they are thicker. Likewise, the thicker portions of the continents not only produce the highest mountains, they also have roots that extend deep into the mantle. The effect is like that of an iceberg with much of its bulk beneath the ocean surface.

Tectonic Processes

In spite of our traditional image of rocks as timeless and unchanging, the outer portion of the earth in fact is in constant motion. Discoveries in the middle 1960s led to a model for earth movements known as *plate tectonics* (tectonics means large scale earth movements). According to the model, the crust and outermost mantle are divided into major sections, or *plates,* which are constantly moving relative to one another around the globe. The plates themselves behave rigidly and show little effect of this motion, but at the plate boundaries intense activity occurs. If two plates are moving toward one another, one plunges beneath the other, eventually being consumed by assimilation deep in the mantle. If two plates are moving apart, magma oozes into the space between them, solidifying as new crust. And if two plates are moving past each other, the friction between them produces great earthquakes.

Plate tectonics actually has its seeds in a theory from early in the century called *continental drift*. The theory stated that prior to 200 million years ago all of the continents had been parts of a single mass named *Pangea*. This supercontinent was thought to have broken up, with separate continental pieces drifting to their present-day locations. The evidence that was cited included remarkable similarities in geology along different coastlines when the continents were figuratively put back together, and distributions of similar fossil types from different, now isolated, continents, particularly those in the southern hemisphere. According to that theory the continents were supposed to be plowing through the rocks of the ocean basins, like a ship through water, with oceanic crust being overridden at the prow of the continent and quietly reemerging at its wake. The mountain chain along western North and South America was due to compression as the continent pushed over the ocean crust. The essential flaw in the theory is that oceanic rocks do not reemerge at the trailing edge of a continent, but move with the continent as part of the same plate of rocks. We also think now that the motions of the plates include the outer 75 or so kilometers (approximately 45 miles) of the mantle, as well as the crust which overlies it. This combination of upper mantle and crust which makes up the plates is called the *lithosphere*. With the exception of local magma bodies it is solid rock. Beneath the lithosphere between 74 and 253 km (45–155 miles) depth is a zone called the *asthenosphere,* in which perhaps a few percent of the mantle is melted. Since the asthenosphere is not as strong as the lithosphere due to the small amount of dispersed magma, this may account for the mobility of the plates above it.

Plate Boundaries

Boundaries of plates which are moving apart produce a feature called a *ridge* or *rise*. These are linear mountain chains found within each of the major ocean basins (Figure 7–2). They are sites of shallow (less than 100 km; 60 mi) earthquakes, higher-than-normal heat flow, and volcanism. Where two plates move apart, a portion of the mantle is melted, producing a basaltic magma that moves upward, intruding as numerous intersecting dikes or extruding as lava. When this basaltic magma solidifies it becomes new ocean crust attached to one or the other of the diverging plates. Since the separation of North and South America from Europe and Africa about 200 million years ago, oceanic crust has been generated along the Mid-Atlantic Ridge as the continents have drifted away from each other, and the Atlantic Ocean has grown wider. Dividing the width of the Atlantic Ocean by the time since its initial formation, one finds that the Mid-Atlantic Ridge has been producing crust along its length at a rate of 2–3 cm (0.8–1.2 inches) per year.

In the Pacific Ocean, plates are separating along the East Pacific Rise. This feature lies to the west of South America and runs northward, eventually ending in the Gulf of California where new oceanic crust is filling the space created by the separation of Baja California from the mainland of Mexico.

When plates move past one another the boundary between them is known as a *transform fault* (a fault is a surface along which rock has broken and moved). The primary characteristics of this type of boundary are numerous earthquakes. A transform plate boundary is expressed at the sur-

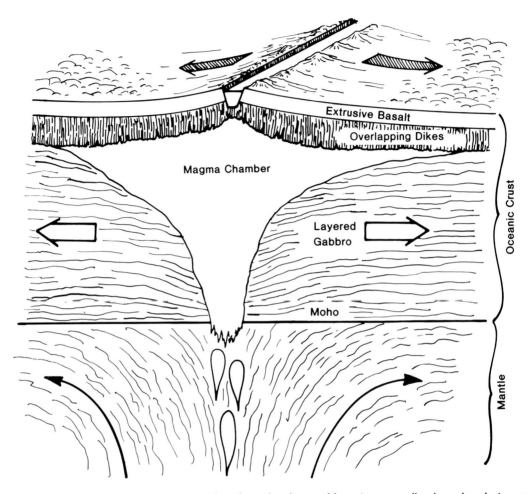

Figure 7–2. Cross-section and perspective view of a rise or ridge at a spreading boundary between plates.

face by a linear zone of numerous subparallel faults. The best-known transform fault is the San Andreas, which begins at the head of the Gulf of California, extends along coastal California, and goes out to sea at Cape Mendocino. Motion along this boundary is about 6 cm (2.4 inches) per year, with the coastal portion moving northward relative to the rest of North America.

The third type of plate boundary, along which plates are converging, is called a *subduction zone*. One plate bends down and is overridden by the other. The lower plate descends at about a 45° angle and is eventually consumed back into the mantle, but not before causing intense activity along the margin of the overriding plate (Figure 7–3). The mountains of western North America owe their existence to the process of subduction along the continental margin, even though, as we will soon see, this process has all but stopped.

The crust of the down-going plate usually is oceanic rather than continental. The juncture of the converging plates is marked by the lowest places on earth, the deep-sea trenches, found, for

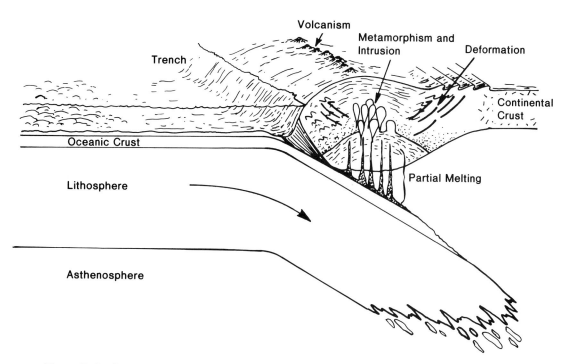

Figure 7–3. Cross-section and perspective view of a subduction zone at a converging plate boundary.

example, along the margin of the western Pacific Ocean and the west coast of South America. As the plate descends it takes with it not only the basalts of the ocean crust, but sediments that have been accumulating on the ocean bottom. These rocks and sediments are heated up due to the rise in temperature with increased depth, as well as friction generated by the grinding together of the converging plates. Eventually melting occurs, in both the crust of the down-going plate and the mantle of the overriding plate. If oceanic crust is overriding oceanic crust, the magmas caused by this melting rise up and erupt as arcuate chains of volcanoes, such as the Aleutian Islands of Alaska. The distinct linearity of island arcs is not coincidental nor is their association with deep-sea trenches. Rather it is a surface reflection of the depth of the subduction at which the critical temperature for melting is reached.

As should be expected, subduction zones are regions of intense earthquake activity, created by fracturing of the rocks as the plates push together. The earthquakes are recorded from depths as great as 600 km (375 miles), much deeper than anywhere else on earth. The reason for this relates to the descending plate. In order to produce an earthquake, rock must be brittle, so that when some critical stress is reached it will break, with the resulting movement along the fracture relieving the stress. If rock is subjected to enough heat and pressure it will behave as a plastic substance rather than a brittle one; that is, it will flow rather than break, and no earthquakes will be produced. Since the down-going plate is cooler than the rock it is plunging into, it will remain brittle and continue to fracture and produce earthquakes to a greater depth than would otherwise be possible.

Mountain Building

The subduction process as described thus far is elegantly simple. However, when a plate descends beneath one which has continental crust at its leading edge, the results suffered by the continental margin are often highly complex. At work are two principal factors, stresses generated by the compression of the plates, and heat transfer from the subduction zone, both by conduction through the rocks as well as by upward movement of hot magmas. The result is a mountain belt in which the rocks have been deformed, metamorphosed and intruded to varying degrees. This process of deformation, metamorphism, and intrusion producing mountains is called *orogeny*. Orogenies tend to be episodic, to have pulses of intensity, and it is customary in geology to name each pulse of activity as a separate orogeny. Arizona, for example, has been affected by the Mazatzal Orogeny and the Laramide Orogeny among others.

Although some generalities can be made about the sequence of events in an orogenic cycle, each mountain range has its own unique history and usually is highly varied along its length.

The stresses produced by compression from the converging plate and by thermal upwellings cause rocks in an orogenic belt to change their shapes, or to deform. The shapes are highly varied in detail, but there are only two main types of deformation: folding and faulting. Which one actually occurs depends on such factors as the rock type, the depth of burial (i.e., the temperature and pressure) and the speed at which the deformation takes place. Folding is the change in shape of a rock without its breaking, whereas faulting is the breaking and movement of rock along fractures. Both types are often found in the same rock. In general, rocks are more brittle close to the surface and become less so with depth, so faulting tends to be more of a surface feature and folding occurs at greater depth. However, there are many exceptions to this statement.

The rate at which a rock deforms is very important. If deformation is slow enough then rocks will become folded even at the surface, whereas if deformation is relatively fast, faulting will occur to great depths. The best analogy for the time dependent nature of rock deformation is Silly Putty. If one pulls slowly on it, it will stretch, but if it is pulled quickly it will snap and break. Rock behaves in a similar fashion, though the time frame obviously is much greater.

Folds appear as wavy or distorted layers in sedimentary or metamorphic rocks (Figure 7–4). Beds or layers that were originally planar become curvi-planar during the folding. The processes that produce folding include compression of rocks caused in a very general way by the convergence of plates, displacement of rocks by the movement of magma bodies, and the sliding and subsequent wrinkling of rocks from a high area to a lower area under the force of gravity.

Folds are identified according to their geometry. A fold which opens downward is called an *anticline*, whereas a fold that opens upward is called a *syncline*. Anticlines and synclines occur together in a folded terrain, in continuity with each other (Figure 7–5). A fold in which strata change from horizontal to steeply dipping, to horizontal again is called a *monocline*. This latter type is relatively rare throughout the world, but many excellent examples are found on the Colorado Plateau in northern Arizona. These will be discussed further in Chapter 12.

Faults are classified according to the relative sense of movement of rocks on opposite sides of the fault plane. *Normal faults* are those on which the rocks above the fault plane have moved downward relative to the rocks below the fault plane (see Figure 7–5). Such a situation produces an extension of the area that is being faulted, and generally occurs when rocks are pulled apart under tension. This type of faulting has been extremely important in the development of the Basin and Range structure and topography in southern Arizona.

Figure 7-4. Folds in Precambrian Mazatzal Quartzite on Four Peaks.

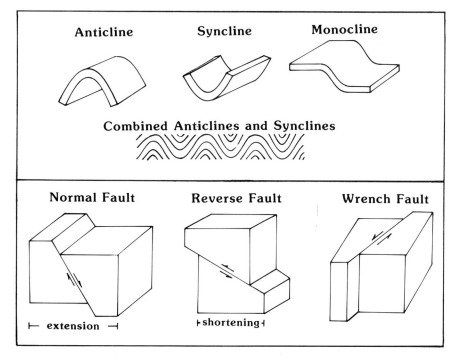

Figure 7-5. Classification of folds and faults.

Reverse faults are those on which the rock above the fault plane moves upward during the faulting. These tend to shorten an area and are generally caused by compression. *Wrench faults* are those in which the motion is horizontal with rock moving neither up nor down. Several wrench faults have been postulated to account for the pronounced differences in rock types between the Colorado Plateau Province of northern Arizona and the Basin and Range Province of western and southern Arizona. When a wrench fault occurs at a plate boundary it is called a transform fault, which has been discussed previously.

Returning to the process of orogeny, a phenomenon of global scale, the following model of an "ideal" mountain belt may serve as a point of comparison for the western United States. The scenario begins with a passive continental margin, with adjacent oceanic crust and continental crust within the same plate. The continental crust consists of mainly old, metamorphic and igneous rocks which have already been through one or more orogenic cycles. Sediments are eroded from the old rocks and deposited in the ocean basin and on the continental shelf, along the continental margin. The sediments may at times accumulate to thicknesses of 9,000–12,000 meters (30,000–40,000 feet), at which depths sedimentary rocks in the bottom of the deposit begin to metamorphose. Such an extensive accumulation of down-warped sediments is called a *geosyncline*.

At some time the continental margin becomes activated, and oceanic crust begins to be subducted beneath the continent. The sedimentary rocks in the geosyncline begin to deform, some are lifted above the ocean surface and become sources of sediments which are deposited in adjacent subsiding basins. The effect at the trench is for sediments and slices of ocean crust to be scraped onto the edge of the continent as the plate descends. When it reaches a critical depth, material taken down the subduction zone will melt and rise, intruding as batholiths or erupting at the surface as volcanoes. As the process continues the rocks deep in the sedimentary pile become progressively more metamorphosed, and partial or complete melting will result. Folding of a very plastic sort occurs in these rocks in the core of the orogenic belt. Often there is upwelling of this soft material, and intrusion into the overlying rock. Rocks nearer to the surface or more toward the continent deform as rigid folds or by thrust faults directed away from the central region. If enough melting occurs magmas will be intruded as large elongate batholiths, parallel to the belt.

While subduction continues, the continent is increasingly heated, causing greater amounts of metamorphism and melting. When the subduction process ends, the region ceases to be compressed. However, magma continues to rise and to be emplaced as plutons or erupt as volcanoes. The region by this time stands very high, the seas have long since retreated and a mountain range has been built. In the absence of compressive stresses the soft and molten portion of the belt may become extended, locally forming downdropped basins in which terrestrial sediments accumulate by erosion from the surrounding mountains. Eventually the region cools, all of the magma solidifies and the mountains are slowly worn down, depositing their sediments at the margin of the continent, where they are ready to begin another orogenic cycle.

Tectonics in Western North America

At the beginning of the Paleozoic Era the area of the western United States was composed of metamorphic and igneous rocks of early to middle Precambrian age, with widespread fault block mountains of tilted late Precambrian sedimentary rocks toward the western margin. In Arizona the older Precambrian had been deformed and intruded between approximately 1.7 and 1.8 billion years ago whereas the younger Precambrian rocks accumulated in shallow seas between approximately 1.5 and 0.8 billion years ago. These include the Grand Canyon Supergroup and the Apache Group. The western limit of the Precambrian crystalline rocks followed a line which trends southwestward from western Montana through western Utah and southern Nevada to southeastern California. An ocean basin lay to the west of Arizona at that time.

Geosynclinal sedimentation occurred throughout the Paleozoic along the continental margin with a thick wedge of sediments, up to 12,000 meters (40,000 feet), accumulating west of the westward facing portion of the continent. Also, seas lapped onto the continent at times intermittently depositing thinner units than in the basin to the west. These rocks are widespread throughout northern Arizona and some formations extend as far south as the Harquahala Mountains in northern Yuma County. The tectonically unstable continental margin along which the above mentioned thick sequence of sediments accumulated from late Precambrian through Paleozoic time, has been named the *Cordilleran Geosyncline*. Paleozoic sediments also accumulated in the southeastern portion of Arizona, probably in a different tectonic basin which has been named the *Sonoran Geosyncline*. There is no evidence of Paleozoic rocks in extreme southwestern Arizona.

The initial events in an area undergoing orogeny are usually not as severe as later developments, which overprint and obscure evidence of the early deformation. Because of this it is often difficult to discern when subduction began in a region. In the western United States there is good evidence of subduction by Jurassic time, but before that the data are meager. Permo-Triassic volcanism, plutonism and faulting appears at scattered localities. However, the earliest record of deformation is found in central Nevada where thrust faulting occurred between the upper Devonian and upper Mississippian. This activity has been called the *Antler Orogeny,* and may represent the earliest Paleozoic subduction along the Pacific margin.

Intense and varied orogenic activity began throughout the western United States in the Jurassic and continued in various forms through the Cenozoic. Along the California coastline deepwater sediments and slivers of ocean crust and mantle were scraped onto the edge of the continent as the ocean plate was subducted at the trench. This very chaotic assemblage of rocks is known as the Franciscan Formation. East of this a continental shelf received sediments from adjacent source areas to the east. Widespread melting occurred in a zone perhaps 100–250 km (62–156 miles) east of the trench. The result was the emplacement of a major belt of intrusive rocks, from granite to diorite in composition, which includes the Idaho Batholith, the Sierra Nevada Batholith of central California and the Coast Range Batholith of southern California. The connection between the Sierra Nevada and Idaho Batholiths was probably somewhat more continuous in the Mesozoic, with separation due to later extensional faulting during the Cenozoic. Plutonism appears to have begun in the Triassic, with peaks of activity in the upper Jurassic and upper Cretaceous. The late Jurassic intrusive maximum and accompanying local deformation has been designated the *Nevadan Orogeny*.

During the Cretaceous, the thick sedimentary rocks that had accumulated to the west of the crystalline continental boundary in southern Idaho, western Utah, and eastern and southern Nevada, were subjected to intense thrust faulting and folding toward the east with the continental block acting as a buttress to their movement. This activity, known as the *Sevier* (pronounced "severe") *Orogeny,* apparently did not extend into Arizona. The rocks were highly metamorphosed in the core of this belt which was uplifted and eroded, with the resulting sediments shed into basins to the west.

During the latter part of the Cretaceous and into the early Tertiary an episode of widespread deformation and plutonism is designated the *Laramide Orogeny*. This produced some thrusting along the margin of the continental block, particularly in southeastern Idaho and western Wyoming. Thrusting in southern and western Arizona is also recorded at this time. However, the predominant deformational style was large, vertical, block uplifts from Montana to Arizona throughout the area underlain by Precambrian crystalline rocks including the Plateau Province of Arizona. Scattered plutonism occurred throughout western and southern Arizona.

The pattern throughout the Mesozoic is one of an orogenic belt expanding to encompass an ever greater area of rock. What had begun in a narrow zone close to the leading edge of the North American plate eventually included all of the continental margin sedimentary rocks as well as the crystalline continental block. The implication is that subduction was continuous throughout this time.

Deformation continued into the Tertiary so that the later stages of Laramide activity merge with succeeding events. In the early to middle Cenozoic, for example, a series of upwellings of metamorphic rocks called "core complexes" occurred in the mobile belt. This was particularly well developed in southern Arizona. Volcanism was also an important process during the Tertiary. In the Eocene, volcanism was largely confined to northern areas, Washington, Oregon, Idaho, Montana and Wyoming. However, in the Oligocene, intermediate to silicic volcanism was widespread throughout the Great Basin, Colorado and southern Arizona and New Mexico. By Miocene this volcanism produced primarily basalt and rhyolite, with andesites becoming relatively scarce.

During the Cenozoic the simple pattern of subduction changed to a more complicated set of plate interactions, which left their own print on the already complex orogenic belt in the western United States. Let us begin 50 million years ago near the beginning of the Cenozoic to follow the series of events (Figure 7–6). An oceanic plate, the Farallon Plate by name, was being subducted beneath the North American plate. Somewhere to the west was a ridge system, the East Pacific Rise, which was generating the ocean crust of the Farallon Plate as well as the crust of the Pacific Plate. In plate tectonics all plates are in motion relative to one another. In order to determine the relative motions it is necessary to hold one plate fixed and observe the motions of the other plates with respect to it. With North America stationary the entire Pacific system, Pacific Plate, East Pacific Rise and Farallon Plate, were moving in a northerly direction. However, the oceanic plates generated at the East Pacific Rise were separating perpendicular to that boundary. When these two motions are combined, the Farallon Plate is seen to have converged toward North America in a northeasterly direction.

As the system evolved, the East Pacific Rise moved closer and closer to the North American Plate. It is important to note that there apparently was a marked offset in the East Pacific Rise by a connecting transform fault. By 30 million years ago the most northeasterly corner of the East Pacific Rise had collided with the subduction zone, somewhere in northern Mexico, and thereafter continued to move into the subduction zone. When the ridge system encountered the subduction zone, the processes of crustal creation met with the processes of crustal destruction, they nullified each other and the generation of new crust ceased in that segment of the plate boundaries. However, the Pacific Plate continued to move northward such that the newly created boundary between it and the North American Plate became a transform fault (the ancestral San Andreas Fault) connecting the East Pacific Rise in the south with the remaining transform fault and rise to the north. This configuration has continued evolving until today. The San Andreas Fault has lengthened as more of the East Pacific Rise has been subducted. About 5 million years ago Baja California began to separate from mainland Mexico along the axis of the East Pacific Rise in that region. The ridge system along the Canadian sector of the Pacific has also been subducted, however a small segment of this ridge system still remains off northern California, Oregon and Washington, producing ocean crust that is still being subducted. The result is that that stretch of coast is the one area in the western United States that is still quite active volcanically.

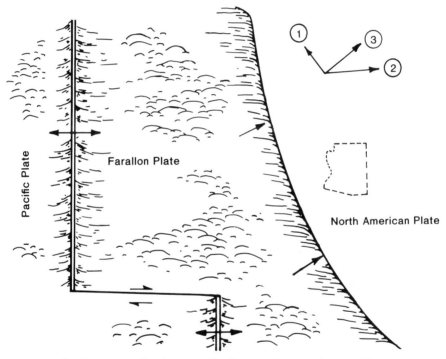

a. 50 m.y. ago—Seafloor spreading between Pacific and Farallon Plates, ridge crest offset by transform fault in lower portion of figure. Subduction of Farallon Plate beneath North American Plates. Arrows indicate plate motions:
(1) Pacific system motion relative to North America,
(2) Farallon Plate motion due to spreading at ridge crest,
(3) resultant of these two motions is the direction of subduction beneath North America.

Figure 7-6. Plate interactions between western North America and oceanic plates during the Cenozoic.

Figure 7–6—Continued.

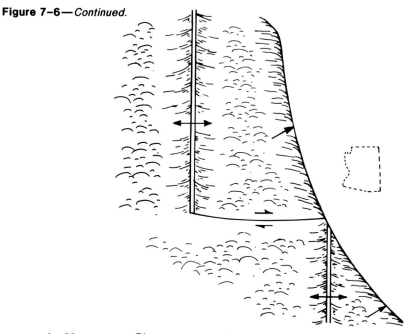

b. 30 m.y. ago—Rise encounters the subduction zone and begins to be subducted.

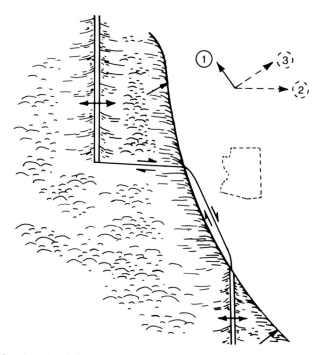

c. 15 m.y. ago—Continued subduction of rise, elimination of motions 2 and 3, leaving only northwesterly motion of Pacific Plate relative to North American plate along the ancestral San Andreas Fault.

Figure 7–6—*Continued.*

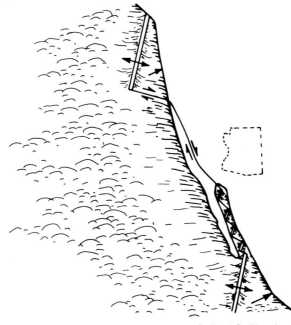

d. Present day—Continued subduction of rise, opening of Gulf of California, further development of San Andreas Fault.

The final major orogenic event in the western United States was the development of the Basin and Range, a region that was extended with the resulting breakup of the crust into a series of blocks bounded by normal faults. Some moved up, others down, creating the characteristic topography of southern Arizona and the Great Basin. The mountain ranges so formed have been rapidly eroded into the adjacent alluvium-filled valleys. Much volcanism has accompanied this event as well.

It is probable that the development of the Basin and Range is due in some way to subduction of the East Pacific Rise, however whether this is due to upwelling and separation of the continent due to action of the ridge beneath the continent or simply due to the relaxation of compressive stresses is not sure. One point worth mentioning is that faulting in the Basin and Range of southern Arizona is now fairly subdued, whereas active separation and accompanying earthquake activity still prevails in the Great Basin.

As we have seen the mountain system of the western United States has had a long and complicated history, owing its origin to the process of subduction at the continental margin. Stages of the activity have occurred in different places at different times. Many of these stages are represented in Arizona, though not all. The foregoing discussion should serve as a background for the geological history of the state which will be discussed in Chapters 9–12.

CHAPTER 8

LANDFORMS

Structural and Physiographic Framework

Arizona sits astride two major geologic provinces, the *Colorado Plateau* to the north and the *Basin and Range* to the south and west. The provinces are characterized by quite different stratigraphic framework and structural patterns, and are separated by a narrow area of transition which is commonly considered a third structural province called the *Transition Zone* or *Central Mountain Province* (Figure 8–1).

The Plateau Province includes the northern third of the state bounded on the south by the Mogollon Rim and on the west by the Grand Wash cliffs in the western Grand Canyon. It extends into southern Utah, southwestern Colorado and northwestern New Mexico. This Province is

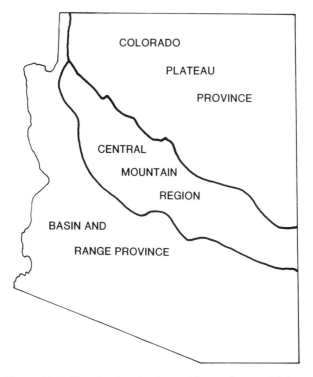

Figure 8–1. Structural—physiographic provinces of Arizona.

characterized by predominantly horizontal stratified sedimentary rocks that have been eroded into numerous canyons, plateaus and scarps along which are exposed many colorful rocks ranging in age from Precambrian to Cenozoic. Many of the most famous landscape features such as the Grand Canyon, Black Mesa, Painted Desert and Petrified Forest, and the Mogollon Rim have been carved into these rocks by erosion. Others such as the San Francisco Mountains and the White Mountains have been piled on top of the Plateau by Cenozoic volcanic activity. These and other landforms are shown on the Physiographic Diagram of the state (Fig. 8–2).

South of the Plateau is a relatively narrow band of landscapes called the Transition Zone or Central Mountain Province, characterized by rugged mountains of igneous, metamorphic and deformed sedimentary and volcanic rocks of Precambrian age, with erosional remnants of Paleozoic age. The elevations are generally lower and the crustal rocks have been more severely faulted than in the Plateau Province. The general absence of Mesozoic and Cenozoic rocks indicates a longer period of erosion and/or nondeposition of sedimentary rocks such as are found in the other provinces. Well known landscape features in the Central Mountain Province include the Black Hills near Jerome and Prescott, the Mazatzal and Sierra Ancha Mountains around Roosevelt Lake, and the Salt River Canyon between Show Low and Globe. The important copper mining districts extending from Jerome to Morenci, and the uranium occurrences in the Precambrian Dripping Spring Quartzite, are located in this area.

The Basin and Range Province includes the southwestern half of the state bounded on the north and east by the Plateau and Central Mountain Provinces along a line trending northwest-southeast from Lake Mead to Globe. It also extends into southern Nevada, southeastern California and into the states of Sonora and Chihuahua. The characteristic landform of this province is one of elongated mountain ranges trending northwest-southeast, separated by broad alluvial valleys. The mountains consist of tilted, and sometimes structurally deformed, blocks of Precambrian, Paleozoic, Mesozoic and Cenozoic rocks that are bounded by faults and have been severely eroded. The Paleozoic rocks are predominantly marine limestones, shales and sandstones that were deposited on a shallow marine shelf in the Early Paleozoic and deeper basins in the Late Paleozoic. The Early Mesozoic rocks are predominantly volcanic or plutonic, and those of Cretaceous age are primarily marine sandstones, shales and carbonates, but also include Laramide intrusives. Cenozoic rocks are largely volcanic but also include nonmarine fluvial and lacustrine sediments, and a small area of marine sediments along the southern Colorado River. The valleys are intermontane depressions that have subsided thousands of feet, and are filled with Cenozoic volcanics, alluvium, and lacustrine sediments. Most of the mountain ranges and valleys in the desert region of western and southern Arizona, including the Hualapai Mountains south of Kingman, the Phoenix Mountains north of Phoenix, and the Chiricahua and Tucson Mountains near Tucson, are examples of Basin and Range landforms.

Evolution of Arizona Landforms

Any area on the earth's surface has a unique combination of landforms that have evolved through time through the effects of physical geologic processes. These processes may be viewed in different ways to help understand and explain why certain landforms are present in an area such as the state of Arizona. One way is to consider all physical processes in terms of whether they are internal or external relative to the earth.

Figure 8-2. Physiographic diagram of Arizona. Reproduced from Smiley and others, Landscapes of Arizona: The Geological Story, *University Press, in press.*

Internal processes include those that operate within the core, mantle or crust and are driven by heat and gravity, often involving dynamic motion of masses of material either in solid form or liquid form. Movement of solid form under high temperatures is generally accomplished by plastic flow very slowly with resultant uplift, subsidence or lateral movement of large volumes of mantle or crustal rock. Under such conditions the rocks at depth are deformed plastically or by chemical reorientation of the component crystals yielding metamorphic rocks. If such movement occurs at lower temperatures but under adequate confining pressure, the resultant deformation is folding without metamorphism or recrystallization. If movements occur so rapidly that the rocks cannot yield plastically then fractures develop structures such as joints or faults.

When the relative displacement of rocks of the crust is upward, the result may be regional uplift or the formation of individual mountain ranges. The uplift of large mountain ranges, accompanied by extensive metamorphism and magma formation is called an orogeny. The uplift of smaller mountain ranges, accompanied by folding, faulting and usually volcanism is generally called a disturbance.

There have been several such tectonic events in the geologic history of Arizona, with much longer periods of quiescence between them. Examples are the Early Precambrian Mazatzal Orogeny (about 1.7–1.8 billion years ago) which formed an extensive mountain chain with a metamorphic and plutonic core across much of Arizona; the Late Precambrian (about 1 billion years ago) Grand Canyon Disturbance which formed numerous smaller block faulted and folded mountain ranges across Arizona. The long times preceding and following each of these diastrophic events were times of subsidence below sea level and slow accumulation of several thousand feet of sedimentary and volcanic deposits. Before the beginning of the Paleozoic these earlier mountains were removed by erosion and regional subsidence. The Paleozoic and Early Mesozoic Eras were times of quiescence and subsidence below or slightly above sea level.

Beginning in late Triassic time and extending through the Jurassic, the Nevadan Orogeny resulted in the uplift, metamorphism and emplacement of granitic plutons in central and southern Arizona, forming the Mogollon Highlands. Most of northern Arizona (the present Colorado Plateau) was a low coastal plain during that time, sloping to a great inland seaway to the north, and to the east and southeast by early and late Cretaceous time.

During the latest Cretaceous and early Tertiary time the whole state was uplifted as a broad arch while the Rocky Mountains to the east in New Mexico, Colorado and northern states were being uplifted during the Laramide Orogeny. Many of the large faults and folds of the Plateau Province (Figure 8–3) were formed at this time. The whole state has remained above sea level since that time.

The latest tectonic event recognized in Arizona is the Basin and Range Disturbance accompanied by faulting and volcanism, resulting in the present topography in the Basin and Range Physiographic Province of southern and western Arizona. The structures formed during that disturbance have been highly modified by late Cenozoic erosion of the highlands and sedimentary filling of the intervening basins. The Mogollon Rim and other portions of the Plateau edge are the result of Cenozoic erosion beginning as early as Oligocene time.

Some of the most prominent landforms in Arizona were formed by another internal process—*volcanism*. This is the result of movement of molten rock material through fractures in the crust

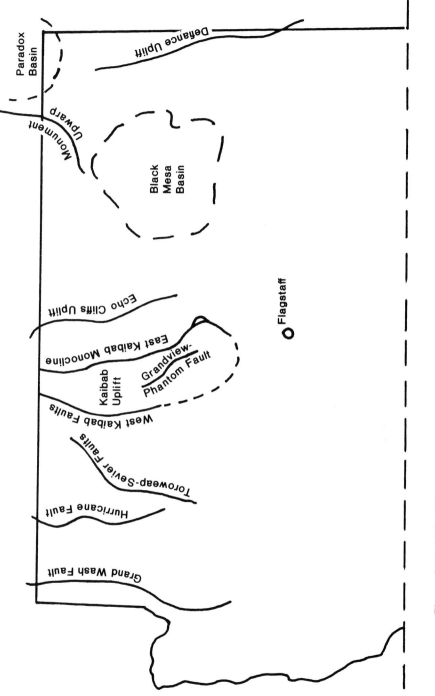

Figure 8-3. Major Tectonic features of the Plateau Province. Modified from Huntoon, 1974; Kelley, 1958; Wilson, 1962.

until it reaches the surface and erupts as lava or pyroclastic material. Volcanism and volcanic rocks are discussed in Chapter 13, therefore they will be discussed here only in the context of their occurrence as landforms in the state. Volcanic landforms are common throughout Arizona, the older ones of middle Tertiary age generally expressed as volcanic necks, plugs and dikes, which are preserved as solidified lava in the vents and conduits of extinct volcanoes (Fig. 13–21). Some familiar examples of such features include the Hopi Buttes, north of Winslow and Holbrook.

Such volcanic remnants commonly occur isolated with no other evidence of the erupted lava or cinders preserved, because they are old enough for erosion to have removed the surface material. Other common volcanic landforms consist of eroded remnants of lava or ash flows, preserved as *mesas* or *buttes*. Mesas are relatively large, flat-topped, elevated regions. Many lava-capped mesas occur in the Central Mountain Region, and portions of the Basin and Range Province, but many fine examples of mesas capped by resistant layers of sedimentary rocks are to be seen in the Plateau Province (Figure 8–4). Buttes are smaller elevated areas with rounded or pointed tops, commonly consisting of tilted fault blocks and capped by lava flow rocks (Figure 8–5). They may be primarily constructional features formed by uplift along faults, or they may be merely erosional remnants of formerly much more extensive features such as mesas (Figure 8–6). With further erosion a butte becomes a *spire* (Figure 8–7).

Some of the more recently formed volcanic land forms in Arizona are relatively unchanged by erosion. The San Francisco volcanic field is the largest, consisting of about 400 distinct vents, mostly composed of piles of cinders, bombs and interbedded lava flows. Many of them, such as Sunset Crater, Merriam Crater and S. P. Crater have well preserved craters at their tops (Figure 13–12). The White Mountain volcanic field is about the same age, as is the Pinacate volcanic field in southern Arizona and northern Mexico.

Obviously not all the mountains and uplifted areas mentioned above are still present. Any such uplifts of the earth's crust are attacked by the *external forces* of weathering and erosion,

Figure 8–4. Padilla Mesa, southwest of Oraibi. The mesa is capped by the Toreva Sandstone and the slope is eroded into the Mancos Shale. Photo by Stephen Trimble, Museum of Northern Arizona.

Figure 8-5. Red Butte, south of entrance to Grand Canyon National Park. It is composed of an erosional remnant of the Moenkopi Formation, capped by Tertiary basalt.

Figure 8-6. Monument Valley, a beautiful display of mesas, buttes and spires left as erosional remnants on the Monument Uplift.

Figure 8–7. A lonely spire of Cow Springs Sandstone at Coalmine Canyon.

resulting in the gradation of the surface by removal of high regions and filling in of the low regions with debris from the highs. These external processes are driven by solar radiation which lifts water into the atmosphere, and gravity which pulls the water and weathered rock material down sloping surfaces. This running water is the primary agent of erosion, which causes the sculpting and ultimate removal of elevated land surfaces.

The lowest elevation to which erosion of land surfaces can occur is sea level, and there must be some gradient down which running water can flow. The most gradual gradient along which a given stream can flow is called its base level, and when it reaches that gradient no more erosion can occur.

Many landforms are erosional features that have been formed by the dissection, due primarily to running water, of elevated topographic features such as plateaus, mesas, or mountain ranges. Most of the well known scenic places in the Plateau Province of northern Arizona are canyons that have been cut down into the elevated plateau—Grand Canyon, Oak Creek Canyon, Canyon DeChelly, Walnut Canyon, Monument Valley and many others.

A progression of landforms evolve in elevated terrain as water drains from it over long periods of time. At first the streams cut narrow, steep-sided canyons with V-shaped bottoms, generally following relatively straight courses. At this stage the streams are widely separated with wide drainage divides between them. This stage in the evolution of a landscape is called "youthful" (Figure 8–8). Much of the landscape in the Plateau Province falls in this category.

Figure 8–8. Canyon DeChelly, a canyon eroded into the Defiance Uplift by a tributary to Chinle Wash. It is an example of a youthful landscape.

As time passes, the stream valleys become wider due to the lateral erosion of valley walls by tributaries and the lateral shifting of the main stream back and forth across its flattened valley floor. At this stage of landscape development, called "maturity," the stream gradient is less steep, therefore its erosive power is reduced. Due to the reduction of erosive and transporting power, and the increase of sediment load brought down by the tributaries, the main stream becomes overloaded with sediments and some of the load settles out within the channel. This "choking" of the channel by beaches, sand or gravel bars, deltas, and so forth, causes the stream to shift laterally and it becomes a meandering stream on a flat aggrading valley floor (Figure 8–9). In the mature stage the stream divides are reduced in area to approximately the same area as the intervening valleys. The stream gradient is much nearer base level (at least a local one) than is a youthful stream, therefore there may be little or no downcutting. In fact the valley may be aggrading or actually gaining in elevation. Several Arizona valleys in which such deposits have been built up in the past are the Safford, Tonto Basin, Verde Valley and Salt River Valley, which contain hundreds to thousands of feet of gravel, sand, silt, mud, limestone and salt deposits.

After a very long time the continued erosion of drainage divides by tributary streams and the aggradation of valley floors by meandering streams will cause the valleys to be several times larger in area than the divides. This stage of landscape evolution is called "old age." There are probably

Figure 8–9. Stream meanders in a drainage system, forming as the stream gradient is reduced and the sediment load is too great for stream transport.

no old age landscapes in Arizona because the whole state has recently been, and probably still is, tectonically active and the stream gradients are too steep to allow such landscapes to be developed.

The drainage system and landscape of an area may be rejuvenated during any stage of evolution, by uplift of the area or lowering of local base level. Base level may be lowered by a change in the drainage pattern due to erosion through a barrier (lava flow, landslide, drainage divide). When rejuvenation occurs, the stream pattern often remains the same, except the canyons become narrow, steep walled and V-shaped again. Many examples of meandering streams with such canyons can be observed on the Colorado Plateau (Figures 8–10 and 14–5). Such streams are *antecedent* if they maintain their courses while the land is uplifted beneath them, or *superposed* if the streams have been "let down" deeper and deeper as successive layers of rock are removed by the erosional processes.

In some instances, streams appear anomalous in that they seem to cross impassable barriers. The best example in Arizona is the Colorado River in the Grand Canyon, where it cuts a 5000–foot deep canyon across the Kaibab Uplift on its way to the Gulf of California, rather than flowing around the uplift (Figure 8–11). John Wesley Powell, the first to run the Colorado River through the Grand Canyon (in 1869), claimed that the river was there first and merely maintained its course as the Kaibab Uplift rose beneath it. Subsequent geologists have proposed that the uplift

Figure 8-10. Incised meanders of Oak Creek near Cornville in the Verde Valley. The stream is becoming incised into the Verde Formation as a result of rejuvenation of the Verde River drainage.

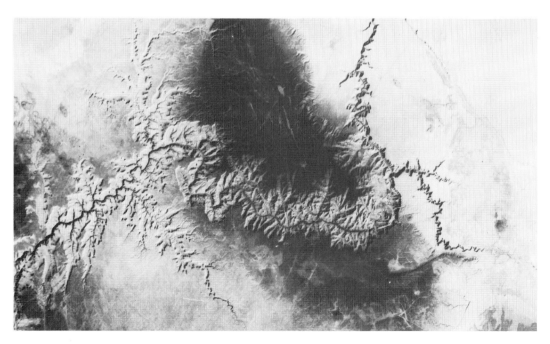

Figure 8-11. The Grand Canyon of the Colorado River. Note how it cuts across the arched surface of the Kaibab Uplift which is visible as an elongate black area on the photograph.

was there first and the Colorado River, which was flowing westward from it, gradually cut its canyon further and further into the uplift by headward erosion until it captured the Little Colorado River that was flowing southward on the eastern side of the uplift. The evidence cited for this latter interpretation includes indications that the flow direction of the Little Colorado River has been reversed (from southward to northward flow) since late Tertiary time. The large area of late Tertiary lacustrine sediments called the Bidahochi Formation in the Hopi Buttes area north of Winslow and Holbrook, is believed to have been the outlet for the ancestral Little Colorado River prior to the reversal (Figure 8–12).

Another argument for this latter theory lies in the fact that the Kaibab Uplift is probably much older (Laramide or Cretaceous-Early Tertiary age) than the Colorado River canyon (Grand Canyon) which was not cut until late Tertiary time (6–8 million years ago). Evidence for the relative youth of the Colorado River includes the occurrence of middle and late Tertiary fluvial and lacustrine sediments in the Colorado River drainage in the western Grand Canyon. These deposits (the Muddy Creek Formation) and absence of Laramide age sediments suggest that the Colorado River was not flowing there until after the uplift occurred. There is some evidence that the Gulf of California extended as far north as the Lake Mead area in middle Tertiary (Miocene)

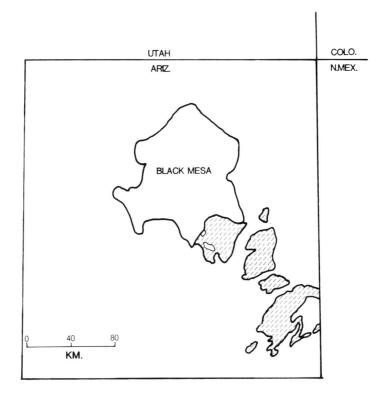

Figure 8–12. Map of area of exposures of the Pliocene Bidahochi Formation (lined pattern). These are remnants of the extensive lake deposits that were continuous across the area during the Pliocene. Modified from Oetking, 1957, geologic highway map of the Southern Rocky Mountain region, American Association of Petroleum Geologists.

time, and it certainly extended north to Needles as late as Pliocene time (Figure 8–13). A precise minimum age for the cutting of the Grand Canyon has been determined at Lava Falls where lava flows from Vulcans Throne (dated at 1.2 million years old) flowed down from the north rim to the present canyon floor (Figure 13–20). This proves that that portion of the canyon was cut prior to 1.2 million years ago. Since then the river has eroded its channel down through about 180 meters (600 feet) of hard basalt, documenting a cutting rate of about 180 meters per million years. Extrapolating that same rate, it can be calculated that the Colorado could have eroded its canyon 900 meters (3000 feet) deep at that point + 180 meters (600 feet) of additional basalt in 6 million years.

Other studies along the southern and western margins of the Colorado Plateau indicate that the plateau has been an elevated region since Oligocene time (50 million years ago) and support the argument that the uplift of it and features on it such as the Kaibab Uplift were there prior to the cutting of the Grand Canyon. During that time at least 2000 meters (6000 feet) of Mesozoic and Cenozoic sedimentary rocks have been eroded from the southern and western portions of the plateau in Arizona and the rim has been eroded back at least fifty miles to its present position.

South and west of the Plateau Province is an elongate, arcuate belt of Precambrian and Paleozoic rocks that have many structural similarities with the Plateau. This belt, called the *Central*

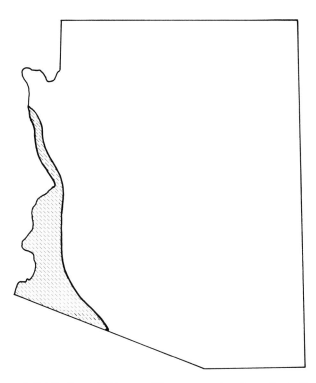

Figure 8–13. Map of distribution of Miocene-Pliocene marine deposits in the ancestral Gulf of California.

Mountain Province, is presently rugged and mountainous with elevations similar to the plateau. Its geologic history indicates that it has been tectonically more stable and has remained above sea level throughout most of the Paleozoic and Mesozoic, at times when the areas of the Plateau Province and southeastern Arizona were usually near or below sea level. This stable region during the Paleozoic Era has been called Mazatzaland and during the Mesozoic has been called the Mogollon Highlands (Figure 11–1). Being an elevated region, it was eroded and served as a source of sediments that were deposited in the low-lying regions to the north, south and west in Arizona. The Central Mountain Province includes many rich ore deposits that have been, and some still are being exploited for minerals such as copper, gold, and silver. Some well known mining districts in this province are Jerome, Bagdad, and Globe-Miami.

To the northwest the Central Mountain region loses its distinctive character near Kingman due to the dominance of Basin and Range structure. To the east it loses character due to Basin and Range faulting and the cover of volcanic rocks in the White Mountains volcanic field.

A third clearly defined structural-physiographic province comprises the landforms of the remaining southern and western portions of the state. This area, called the *Basin and Range Province* is characterized by numerous, elongate block-fault mountain ranges (horsts), separated by wide, flat alluvium-filled valley floors (Figure 8–2). The valleys are mostly deep sedimentary basins (grabens) formed by the downward movement of one or more high-angle normal faults. Subsurface studies of several of these basins have shown that they are filled with thousands of feet of alluvial debris (gravel, sand, silt) that was eroded from adjacent mountain ranges. Some have thick sequences of evaporite deposits (salt, gypsum, etc.), indicating long periods of interior drainage in which water accumulated and evaporated leaving layer upon layer of salts to accumulate on the rapidly subsiding basin floors, interbedded with clastic debris being washed in from the mountains.

Several types of constructional landforms are typical of the boundary between basin and range mountains and valleys. Alluvial fans are delta-shaped masses of sand, gravel and silt that are carried down intermittent stream channels in arid mountains to the margin of a relatively flat valley floor where the stream loses its transporting energy abruptly and its sedimentary load is dropped. The result is a gently sloping pile of roughly stratified material called an *alluvial fan* (Figure 8–14). Where several such alluvial fans coalesce laterally along a mountain front the sloping surface is called a *bajada* (Figure 8–15).

Where erosion is the dominant process along a mountain front, a sloping bedrock surface may develop. This surface which may be bare bedrock or may be covered by a thin layer of sediments, is called a *pediment* (Figure 8–16). Such bedrock pediments may extend some distance from the mountain front toward the valley, where its edge marks the position of a bounding fault.

Figure 8-14. Alluvial fans at north end of Tonto Basin.

Figure 8-15. Bajada, formed by coalescing of alluvial fans along foot of Peacock Mountains, east of Kingman.

Figure 8–16. Pediment surface eroded into granite north of Kitt Peak.

PART II

GEOLOGIC HISTORY OF ARIZONA

INTRODUCTION

The earth has continuously evolved since its beginning, eons ago. Mountain chains have risen and been reduced to level plains. The oceans have many times encroached onto the continents, and then retreated to the confines of the ocean basins. Ocean basins themselves have been created and others have disappeared between converging continents. The record of the ocean basins prior to the Mesozoic is largely lost, but the history of the continents remains incompletely preserved in the rocks exposed throughout the world. Different regions experienced different sequences of events, and early events are often overprinted or obliterated by later geologic activity.

In Arizona the Cenozoic activity in the Basin and Range Province has caused many of the rocks to be buried by alluvium in deep basins, and on the Colorado Plateau the Precambrian rocks are for the most part hidden beneath a thick cover of Paleozoic and Mesozoic sediments. The record is fragmentary; some periods are better preserved and therefore better understood than others. When the evidence is limited we must strain to decipher the grand events that took place.

Arizona's geological history is particularly long and full. The record begins in the Precambrian, about 2 billion years ago. In some parts of the world, rocks created during the Precambrian have been unaffected during later geological periods, but Arizona has been the site of either active tectonism or deposition almost continuously from its beginning to the present day. The areas of exposures of these rocks are shown on the generalized geologic map of Arizona (Plate I).

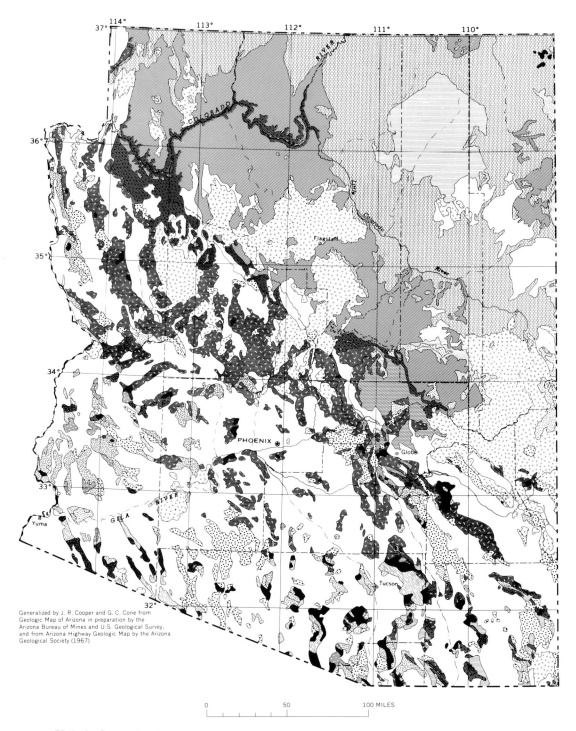

Plate I. Generalized geologic map of Arizona. From Arizona Bureau of Mines, Bulletin 180. Reproduced by permission of publisher.

E X P L A N A T I O N

SEDIMENTARY AND VOLCANIC ROCKS

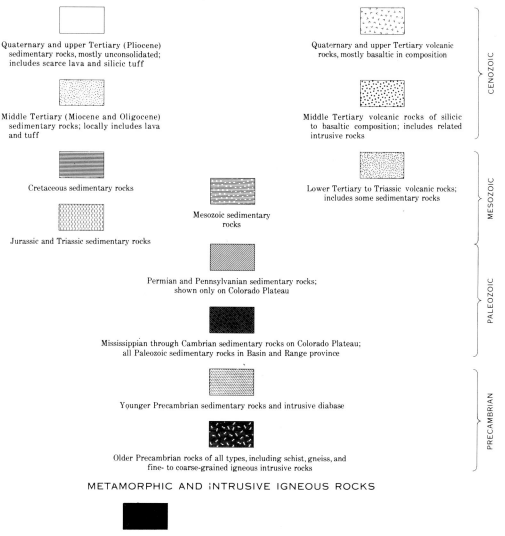

Quaternary and upper Tertiary (Pliocene) sedimentary rocks, mostly unconsolidated; includes scarce lava and silicic tuff

Middle Tertiary (Miocene and Oligocene) sedimentary rocks; locally includes lava and tuff

Cretaceous sedimentary rocks

Jurassic and Triassic sedimentary rocks

Mesozoic sedimentary rocks

Permian and Pennsylvanian sedimentary rocks; shown only on Colorado Plateau

Mississippian through Cambrian sedimentary rocks on Colorado Plateau; all Paleozoic sedimentary rocks in Basin and Range province

Younger Precambrian sedimentary rocks and intrusive diabase

Older Precambrian rocks of all types, including schist, gneiss, and fine- to coarse-grained igneous intrusive rocks

Quaternary and upper Tertiary volcanic rocks, mostly basaltic in composition

Middle Tertiary volcanic rocks of silicic to basaltic composition; includes related intrusive rocks

Lower Tertiary to Triassic volcanic rocks; includes some sedimentary rocks

CENOZOIC

MESOZOIC

PALEOZOIC

PRECAMBRIAN

METAMORPHIC AND INTRUSIVE IGNEOUS ROCKS

Tertiary and Upper Cretaceous intrusive igneous rocks

Post-Paleozoic gneiss and schist

Mid-Cretaceous to Triassic intrusive igneous rocks

CHAPTER 9

THE PRECAMBRIAN ERA

Two major groups of Precambrian rocks occur in Arizona. For convenience these are called the Older Precambrian and the Younger Precambrian.

The Older Precambrian rocks are exposed at the bottom of the Grand Canyon, at numerous localities in the mountainous region south of the Colorado Plateau, and scattered throughout many of the ranges in the Basin and Range (Figure 9–1). The older group of Precambrian rocks is arranged into two provinces with closely-spaced but distinct ages. The boundary between the two provinces runs northeast-southwest in the general area of the Mazatzal Mountains and upper Tonto Basin.

Northwest of this boundary the rocks belong to the Yavapai Series south of the plateau and to the Vishnu Schist in the Grand Canyon. Volcanic rocks, mainly of intermediate composition, dominate the sequence, but volcanogenic and clastic sedimentary rocks occur as well. This sequence is similar to other rocks of approximately the same age found on every continent, but debate exists as to whether they originated in some sort of island arc at the margin of a continent or in a region where continental blocks were separating. The rock units discussed in the following pages and their approximate correlation are shown in Table 9–1.

The oldest rocks in the Grand Canyon have been mapped and given the names Vishnu Schist, Trinity Gneiss and Elves Chasm Gneiss, all of which were originally sedimentary rocks and volcanic sediments and flows. These rocks were deformed, metamorphosed and intruded by the Zoroaster Granite about 1.7 billion years ago resulting in a mountain range as great as the Rockies are today.

Different stratigraphic units have been identified locally in the Yavapai Series in areas that are separated by large areas of granitic plutons so that continuity of outcrops is lacking, and correlation between them is uncertain. In the Bagdad area, for instance, the youngest formation is the Hillside Mica Schist derived originally from sandstone and shale. This overlies the Butte Falls Tuff, a water-laid volcanic ash. A third unit, the Bridle Formation is thought to underlie the Butte Falls Tuff, but the relationship is unclear. This rock is metamorphosed andesite and basalt flows, interbedded with water-laid tuff and sedimentary rocks. A total of more than 2700 meters (9000 feet) of metavolcanic and metasedimentary rock occurs.

In the area included by Prescott, Jerome and Black Canyon City two different groups of the Yavapai Series are recognized: the Ash Creek Group and the Big Bug Group. The groups are separated by a prominent north-south linear feature known as the Shylock Fault Zone, which runs parallel to Interstate 17 at the eastern end of the Bradshaw Mountains and on north into the western side of Mingus Mountain. The Ash Creek Group is to the east and Big Bug Group is to the west, although some of the Big Bug Group also occurs east of the Shylock Fault Zone around

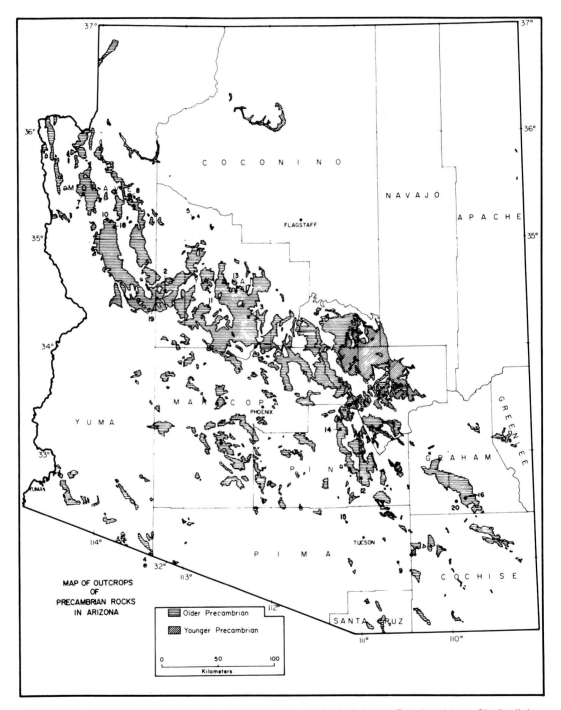

Figure 9-1. Map of outcrop locations of Precambrian rocks in Arizona. Reprinted from Shafiqullah, M., and others, 1980, K/Ar Geochronology and geologic history of southwestern Arizona and adjacent areas: *Arizona Geological Society Digest* 12.

Black Canyon City. The relationship between the two groups is uncertain, but it is thought that some if not all of the Big Bug Group is younger than the Ash Creek Group.

The Ash Creek Group includes, from youngest to oldest, the Grapevine Gulch Formation, composed of breccia, tuff and dacite flows, the Deception Rhyolite, the Brindle Pup Andesite, the Buzzard Rhyolite and the Gaddes Basalt. The combined thickness of formations totals around 6000 meters (20,000 feet).

The Big Bug Group consists of three formations. The youngest, the Iron King Volcanics, is predominantly andesite and basalt flows with some rhyolite tuffs interbedded locally. The Spud Mountain Volcanics are volcaniclastic in their lower part and grade upward into andesitic tuffs. The oldest formation, the Green Gulch Volcanics, contains minor conglomerate at its base, and is overlain by basalt and rhyolite flows. The Big Bug Group is foliated and severely folded in the Bradshaw Mountains, but less deformed east of the Shylock Fault and to the west in the Prescott area.

Radiometric dating done largely by L. T. Silver and co-workers at the California Institute of Technology indicates that volcanic rocks of the Yavapai Series and Vishnu Schist were extruded between 1750 and 1820 million years ago. The rocks were subsequently deformed and intruded by magmas of mainly granodiorite composition, but ranging to gabbro in some cases. These include the Zoroaster Granite in the Grand Canyon and the Government Canyon and Brady Butte Granodiorites south of Prescott. The plutons invaded extensive areas, between isolated occurrences of metamorphic rocks. Some of these plutons were emplaced before deformation had stopped, but most are post-tectonic. Radiometric dates on these intrusive rocks range between 1725 and 1770 million years in age.

In the Mazatzal Mountains and the area northeast of the Tonto Basin is a suite of rocks dominated by volcanics or volcanically derived sedimentary rocks. These rocks are thought to be about the same age as the Pinal Schist, which crops out in the Little Dragoon and the Pinal Mountains. Together these two groups of rocks represent a nearshore facies in the Tonto Basin area and deeper water facies to the southeast. The rocks apparently are younger than the Yavapai Series, exposed to the west and northwest, which may have been an eroding source area during accumulation of the sediments to the southeast. Nowhere have the Yavapai Series and the sequence of Precambrian rocks in the Mazatzal Mountains been seen in contact, so radiometric dating must be used to discern the relative ages of the two provinces. Several of the volcanic units give dates of 1715 to 1730 million years, coincident with the cessation of major intrusions into the Yavapai Series, by which time that province would have been uplifted and erosion begun.

The stratigraphic sequences in the Mazatzal Mountains and the Tonto Basin have been worked out separately, but they are similar enough that correlations are possible. In both areas the lowest group of rocks have been called the Alder Group. Volcanic rocks, an important constituent in both areas, range in composition from rhyolite to basalt, but are predominantly of the silicic types. Volcanically derived sediments, conglomerate, sandstone and shale also occur. In the Mazatzal Mountains the Alder Group is divided into, from youngest to oldest, crystallithic tuffs, the West Fork Formation, the Horse Camp Formation, the Cornucopia Formation, the East Fork Formation, the Oneida Formation and the Telephone Canyon Formation. This sequence is in excess of 4200 meters (13,860 feet). In the Tonto Basin the units include pre-Breadpan Rocks, the Breadpan Formation, the Flying W Formation, the Houden Formation and the Board Cabin Formation. These rocks total more than 3800 meters (12,540 feet) thickness. Although similarities exist between the two localities of Alder Group, the individual formations have not been correlated.

Geologic correlation chart of Precambrian rocks of Arizona.

Grand Canyon	Baghdad	Prescott / Jerome / Black Canyon City	Mazatzal Mountains	Tonto Basin and Salt River Canyon	Southeastern Arizona
		West ← → East			
Chuar Group					
Sixtymile Formation					
Kwagunt Formation					
Galeros Formation					
Nankoweap Formation					
Unkar Group					
Cardenas Lavas — 1.09 B.Y.				Troy Quartzite	Troy Quartzite
Dox Sandstone				**Apache Group**	**Apache Group**
Shinumo Quartzite				Basalt	
Hakatai Shale				Mescal Limestone	
Bass Limestone				Dripping Spring Quartzite	Dripping Spring Quartzite
				Pioneer Shale	Pioneer Shale
			Mazatzal Quartzite	Mazatzal Quartzite	Pinal Schist
			Maverick Shale	Haigler / Payson	Johnny Lyon Granodiorite 1.55–1.65 B.Y.
			Deadman Quartzite		

Top band: Younger Precambrian ←——→ | Diabase 1.1 B.Y. | [Old]er Precambrian

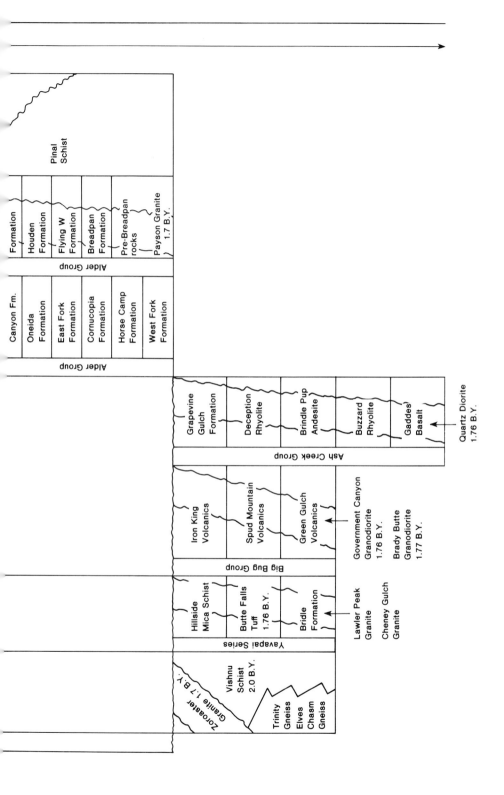

Overlying the Alder Group in the Mazatzal Mountains is the Red Rock Rhyolite. The formation is more than 1200 meters (3960 feet) thick, and includes welded tuffs, lavas, breccias and massive, featureless varieties of rhyolite. The Haigler Rhyolite of the Tonto Basin area is correlative. There the unit is about 2000 meters (6600 feet) thick, is similar in composition, but also includes minor amounts of basalt, conglomerate and other clastic rocks. When viewed as a whole this volcanic deposit is as large as the largest known silicic volcanic fields, such as the San Juan Mountains in southwestern Colorado. The chemistry of these rhyolites is particularly enriched in potassium and sodium. The Haigler Rhyolite includes numerous intrusive phases, and is also intruded by the contemporaneous Payson Granite, which has a similar composition.

The volcanic sequence is overlain by clastic rocks which probably accumulated as the ocean transgressed the volcanic piles and erosion commenced. In the Mazatzal Mountains three formations are recognized, the lowest is the Deadman Quartzite, followed by the Maverick Shale, followed by the Mazatzal Quartzite totaling more than 660 meters (2180 feet) thickness. In the Tonto Basin shales are not recognized and all rocks are assigned to the Mazatzal Quartzite which is more than 800 meters (2640 feet) thick. Basal conglomerate occurs at some places with large clasts of the underlying rhyolite. The quartzites are usually purplish in coloration; they are fairly clean and cross bedded, and in general are very tough rocks.

The Mazatzal Quartzite was so resistant to erosion that it remained as islands during the deposition of the Younger Precambrian, as well as the Paleozoic sediments in the Payson area. At two places, Christopher Mountain near Kohl's Ranch, and west of Tonto Natural Bridge, the Paleozoic sedimentary rocks pinch out against highlands of the Mazatzal Quartzite. At the southern end of the Mazatzal Mountains, Four Peaks, which is also capped by Mazatzal Quartzite, stands high and jagged above the surrounding area, composed of granite which has weathered to more subdued forms (Figure 9–2).

Activity designated the Mazatzal Orogeny ensued after deposition of the Mazatzal Quartzite approximately 1650 to 1730 million years ago. The thick volcanic units prevented much small-scale deformation, except in some of the sedimentary units. Compression first caused open folding with axes trending generally in a northeast-southwest direction. This was followed by thrust faulting toward the northwest which is exposed in the northern Mazatzal Mountains. This was in turn followed by normal faulting, perhaps strike-slip faulting and intrusion of large volumes of post-tectonic granite. These plutonic rocks are found in the southern Mazatzal Mountains (they are well displayed along the Beeline Highway, route 87 between Mesa and Payson) and in many places to the southeast associated with the Pinal Schist which was also affected by the Mazatzal Orogeny. Dating on the post-tectonic granites ranges down to about 1650 million years, marking the end of the orogeny.

The Older Precambrian rocks in the southeast portion of the state are assigned to the Pinal Schist which crops out in the Globe-Miami area, the Pinal Mountains, the Little Dragoon Mountains, the Mule Mountains and in a few other scattered locations. They are apparently synchronous with the rocks in the Mazatzal Mountains—Tonto Basin area. These rocks also were deformed and intruded during the Mazatzal Orogeny.

Prior to metamorphism these rocks were mostly graywackes (dirty sandstones) and shales, characteristic of deep-water marine conditions like those found at a continental margin. The

Figure 9-2. Four Peaks at southern end of the Mazatzal Mountains. Crest composed of the Mazatzal Quartzite, and lower slopes of Precambrian granite.

sequence is similar throughout and so has not been subdivided. A few minor volcanic units of mafic and silicic composition occur at places. A thickness of up to 6000 meters (20,000 feet) has been estimated in the Little Dragoon Mountains.

The great mountain ranges formed during the Mazatzal Orogeny were worn down over the next half billion years to a nearly flat and featureless plain (similar to the Canadian Shield of eastern Canada), which then subsided beneath a shallow marine environment on the continental shelf. The sediments that accumulated (3,000+ meters; 10,000+ feet) in and near (on shore, deltas and tidal flats) the shallow marine environment during the next half billion years (the Late Precambrian), varied from volcanic rocks to sandstones to limestone and evaporites, and are well-preserved and relatively undeformed. Several distinct formations are recognized which have been lumped together as the Grand Canyon Supergroup (including the Unkar and Chuar Groups) in the Grand Canyon, and Apache Group in central and southeastern Arizona. All of the younger Precambrian strata were disturbed by normal faulting and folding and some volcanic intrusions of dikes and sills. This period of deformation has been named the Grand Canyon Disturbance, which resulted in the formation of NW-SE trending Basin and Range type mountains. The formations of Younger Precambrian age are shown in the correlation chart (Table 9–1).

The oldest fossils in Arizona are found in sedimentary rocks between 1.5 and 1.0 billion years old. Such rocks are found in the Grand Canyon, Salt River Canyon, Sierra Ancha Mountains, and elsewhere and include single-celled, blue-green algae which caused, during life, the precipitation of thin limestone laminae within the algal tissues. These fossils are called stromatolites and are common in the Bass Limestone of the Grand Canyon and the Mescal Limestone of the Salt River Canyon and Roosevelt Lake areas (Figure 9–3 and 9–4). Microscopic cellular plant fossils (algae) have been discovered recently in the Kwagunt Formation of the Grand Canyon.

Stromatolites are still forming today in several parts of the world, primarily in supratidal environments, generally associated with evaporite minerals (salt, dolomite and gypsum). Salt crystal casts, mud cracks and evaporites are also common in the Bass Limestone, suggesting a similar environment of deposition in northern and central Arizona about one billion years ago.

The upper Precambrian age of the Bass and Mescal is determined by superposition and cross-cutting relationships, that is, it is beneath sedimentary rocks containing trilobites and other Cambrian fossils, and is cut by radiometrically dated igneous dikes (1.0 billion years old) that penetrate the Bass Limestone and all other formations below the Cardenas Lavas. The fossils in the Kwagunt Formation lie above the Bass Limestone, therefore are younger, but are also beneath the Cambrian age rocks, making them Precambrian in age. Most of the other younger Precambrian sedimentary rocks are sandstones or oxidized shales that formed in environments such as deltas, tidal flats or intertidal zones in which delicate organisms could not be preserved as fossils. A few other reports of fossil jellyfish, mollusks and trilobites in Precambrian rocks of the Grand Canyon have been generally discredited.

Figure 9–3. Steeply dipping Mescal Limestone at Roosevelt Dam. The dam was constructed of blocks of the limestone.

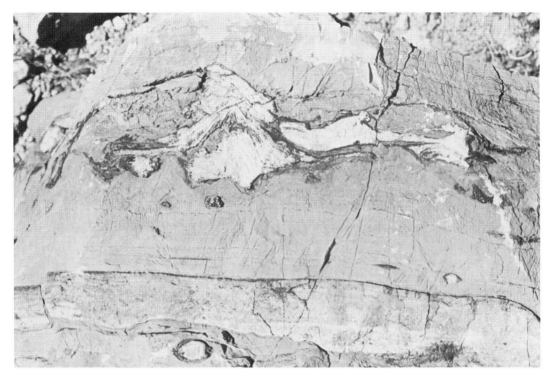

Figure 9-4. Algal stromatolites in the Mescal Limestone at Roosevelt Dam.

CHAPTER 10

THE PALEOZOIC ERA

The Paleozoic and Mesozoic rocks in Arizona contain a variety of stratigraphic units and fossils that yield information on the conditions which prevailed at the time those rocks were formed. They tell us that during much of this portion of Arizona's geologic history, it was covered partly or entirely by ocean water, and the organisms that lived in the water or on the sea floor were buried in sand or mud and preserved as fossils. At other times the area was above sea level, either being eroded or covered by even layers of river-borne mud or wind-blown sand. These sediments sometimes buried plants or terrestrial animals and they were preserved as fossils. In this chapter we will discuss the probable environments of deposition of sedimentary rock units of Paleozoic age based on the physical and biological criteria contained within them.

Also, as explained in Chapter 4, fossils tell us the relative age of the rocks in which they were formed. To the trained eye, a fossiliferous outcrop can quickly be correlated with a standard chronology established on the basis of superposition of strata and fossil succession, and the age of the rock determined (Table 4–1). Maps of the state showing the area covered by rocks of a given period of time are included for most of the geologic periods in chapter 10.

Cambrian (600–500 Million Years Ago)

Most of the state was covered by shallow seas that encroached over the area during the Cambrian Period. Marine sediments and fossils accumulated in widespread sheets of sand and mud that apparently covered all the state except for the Defiance Positive area.

The oldest sedimentary rocks that were formed during the Cambrian Period in Arizona typically consist of a basal sandstone unit that rests directly on an eroded surface of Precambrian rock. This contact was named the Great Unconformity by Powell (1875) (Figure 10–1). In the Grand Canyon, western and central Arizona, the basal Cambrian unit is called the Tapeats Sandstone and is lower to middle Cambrian age. Overlying the Tapeats in the Grand Canyon area is the Bright Angel Shale of middle Cambrian age, composed of clay and silt with a characteristic greenish color. It typically erodes more readily than the overlying rocks, forming a broad topographic bench in the Grand Canyon just above the Inner Gorge called the Tonto Platform. The Muav Limestone of middle Cambrian age overlies the Bright Angel Shale and forms a steep slope broken by low cliffs at the top of the Tonto Platform. The Muav Limestone consists of interbedded limestones and greenish shales similar to the Bright Angel Shale, the two formations being gradational into each other, meaning they do not have a distinct, clearly defined boundary. The Tapeats Sandstone, Bright Angel Shale and Muav Limestone compose a recognizable stratigraphic unit called the Tonto Group (Table 10–1). Fossils in the Cambrian of the Grand Canyon consist

Figure 10-1. The Great Unconformity. Cambrian Tapeats Sandstone resting on an erosional surface on Older Precambrian Vishnu Schist in Grand Canyon.

of a variety of trilobites and some inarticulate brachiopods, which have been used by McKee and Resser (1945) to define time-lines for correlation.

In the Virgin Mountains west of the Grand Canyon, more than 600 meters (1,980 feet) of Cambrian marine sedimentary rocks have been divided into several formations that are correlated with those in the Grand Canyon as shown in Table 10-1.

In southern Arizona the basal Cambrian rocks are similar to the Tapeats Sandstone but are given different names—the Bolsa Quartzite (265 meters, 875 feet thick) and the Coronado Quartzite, both of which are middle Cambrian age. Overlying them is the Abrigo Limestone (254 meters; 838 feet thick) of middle to upper Cambrian age, and the El Paso Limestone of upper Cambrian to Ordovician age. These rock units contain fossils similar to those in the Tonto Group in the Grand Canyon, and therefore are the same age.

Cambrian rocks in Arizona record a classic example of a marine transgressive stratigraphic sequence (as discussed in Chapter 3, Figure 3-1) with nearshore sands at the base, followed by deeper water silts and muds, then offshore carbonates at the top of the Cambrian section. They record a slow transgression of the sea from west to east (in the Cordilleran Geosyncline) and south to north (in the Sonoran Geosyncline) across a broad, low-lying erosional surface of low relief, that had been carved into Precambrian rocks of varying age and lithology. The total thickness of Cambrian rocks in Arizona is shown in Figure 10-2. The Defiance Positive was apparently above sea level during the Cambrian Period.

Ordovician (500-440 Million Years Ago)

During the Ordovician, most of Arizona was above sea level, based upon the fact that rocks of that age are found only in the extreme southeastern and northwestern corners of the state. In southeastern Arizona the Longfellow and upper El Paso Limestones (120 meters; 396 feet) contain

TABLE 10–1.   Lower Paleozoic Rock Units in Arizona.

	Virgin Mountains	Grand Canyon	Southeastern Arizona
Devonian	Muddy Peak Limestone	Temple Butte Ls. / Martin Formation	Martin Formation / Portal and Swisshelm Formations
Ordovician	Pogonip Limestone	No Ordovician Recognized	Longfellow Limestone
	Undifferentiated Dolomitic Limestone	Undifferentiated Dolomite	El Paso Limestone
Cambrian	Peasley Limestone	Muav Limestone (Tonto Group)	
	Chisholm Shale	Bright Angel Shale (Tonto Group)	Abrigo Limestone
	Lyndon Limestone		
	Pioche Shale		
	Prospect Mountain Quartzite	Tapeats Sandstone (Tonto Group)	Bolsa Quartzite

This table is intended to show formation names and systemic ages, but not necessarily precise correlations between formations.

Ordovician marine fossils (trilobites, brachiopods, etc.) and in the Virgin Mountains of north-western Arizona the Pogonip Limestone (65 meters; 215 feet) is recognized as Ordovician age (Table 10–1). It is possible that more extensive Ordovician deposits were formed, then removed by subsequent periods of erosion, but in either case rocks of this age are very limited in Arizona.

Silurian (440–400 Million Years Ago)

No Silurian age rocks have been recognized in Arizona. This means they were not deposited because the area was above sea level and undergoing erosion, or any rocks formed then were later removed by erosion prior to Devonian time.

Devonian (400–350 Million Years Ago)

Fossils of Devonian age are marine throughout most of Arizona, indicating a return to shallow marine conditions in most of the state. However, one locality in Salt River Canyon contains fossils of primitive land plants in the Middle Devonian Becker's Butte Member of the Martin Formation, indicating a depositional environment above sea level (Figure 10–3). This plant locality is especially significant because only a few other such fossil occurrences of Devonian age are known in the world.

Such terrestrial environment indicators plus the existence of numerous other localities where Mississippian or Pennsylvanian rocks directly overlie Cambrian or Precambrian rocks indicate a

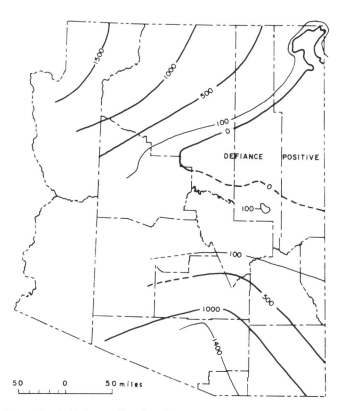

Figure 10-2. Generalized thickness (feet) of Cambrian rocks in Arizona. Reprinted from Peirce, H. W., 1976, Elements of Paleozoic Tectonics in Arizona: *Arizona Geological Society Digest* 10.

long period of erosion and/or nondeposition during or after Devonian time. This erosional surface has a topographic relief of possibly 360 meters (1,188 feet) in the Globe area.

Marine sedimentary rocks, mostly limestone and dolomite, containing abundant fossils, have been assigned to several formations in Arizona. Included are: the Martin Formation (Middle-Upper Devonian) in central Arizona; the Temple Butte Limestone (Upper? Devonian) in the Grand Canyon; the Muddy Peak Limestone (Upper Devonian) in the Virgin Mountains; the Martin Formation (Middle-Upper Devonian) in central and southern Arizona; and the Portal Formation (Upper Devonian) in southern Arizona (Table 10–2).

A limited amount of oil and gas have been produced from Devonian age rocks in northeastern Arizona, and rocks of this age are considered favorable exploration targets where they are buried by younger rocks.

Devonian—Mississippian Disconformity

Throughout most of Arizona a well developed erosional surface exists between the Mississippian marine sediments and Devonian or older rocks. This clearly indicates a time of broad uplift of the state above sea level during which erosion and/or nondeposition prevailed. In the eastern Grand Canyon exposures of the Temple Butte Limestone are restricted to local remnants of chan-

Figure 10–3. Beckers Butte Member of Martin Formation in Salt River Canyon. This is a Devonian land plant locality.

nel fill deposits of sandy dolomite that, due to their protected positions, were not completely removed by the broad scale erosion (Figure 10–4). In western Grand Canyon and central and southern Arizona, Devonian sedimentary rocks of marine origin are generally present, therefore those areas remained below sea level throughout most of the Devonian Period (Figure 10–5). Figure 10–6 shows the thickness of Devonian rocks across the state and tectonic features such as the Kaibab and Defiance Positive areas and the Oraibi Trough. However, by the end of Devonian time, the whole state had been uplifted above sea level and eroded creating an erosional surface that cut into Devonian and older rocks. By early to middle Mississippian time the state was again submerged beneath shallow seas and marine limestones of that age (for example, Redwall Limestone and Escabrosa Limestone) were deposited over the eroded surface creating a disconformity (Figure 3–5).

Mississippian (350–325 Million Years Ago)

Following the general post-Devonian withdrawal of the seas from Arizona and erosion of the Devonian rocks at the surface, there was a general inundation of the area beneath shallow shelf seas. The fossil record in rocks of Mississippian age includes a profusion of invertebrate animals (Figure 5–11) which indicates tropical climatic conditions and shallow, clear water of normal salinity. Under such conditions, calcium carbonate accumulates slowly on the sea floor forming

Figure 10–4. Channel deposit of Devonian Temple Butte Limestone in Grand Canyon. Massive limestones of Mississippian Redwall Limestone above and Cambrian Muav Limestone beneath the Temple Butte Limestone.

Figure 10–5. Devonian Martin Formation forming steep, ledgy slopes beneath cliff forming Redwall Limestone at top. Exposure in Peach Springs Canyon. Photo by Stanley Beus, Northern Arizona University.

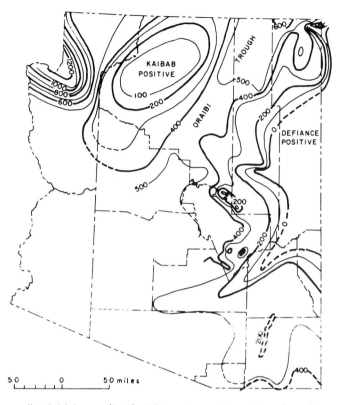

Figure 10-6. Generalized thickness (feet) of Devonian rocks and location of tectonic features. Reprinted from Peirce, H. W., 1976, Elements of Paleozoic Tectonics in Arizona: *Arizona Geological Society Digest* 10.

layers of lime mud and skeletal debris which, through compaction and cementation over a long period of time, forms limestone.

Fossiliferous limestone of Mississippian age is widespread across the state and has been assigned to several formations. The lower to middle Mississippian Redwall Limestone is a prominent cliff former throughout the Grand Canyon and other places in northwestern and central Arizona. It is generally 130–180 meters (429–594 feet) thick, contains varying amounts of chert, and commonly contains caverns due to groundwater solution (Figure 10–7).

In southern and southeastern Arizona, several formations of Mississippian age are recognizable and mappable. These are the Modoc and Tule Spring Limestones in the Clifton area; the Escabrosa Limestone (similar to Redwall) exposed in many of the southeastern Arizona mountain ranges; and the upper Mississippian Paradise Formation in the Chiracahua Mountains. The total thickness of Mississippian rocks is shown in Figure 10–8.

Figure 10-7. Cliff-forming Redwall Limestone at river level in Grand Canyon. Note caverns and flowing spring in cliff.

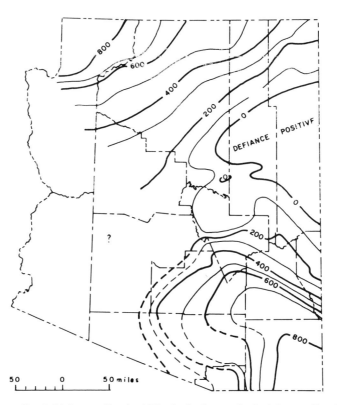

Figure 10-8. Generalized thickness (feet) of Mississippian rocks in Arizona. Reprinted from Peirce, H. W., 1976, Elements of Paleozoic Tectonics in Arizona: *Arizona Geological Society Digest* 10.

TABLE 10–2. Upper Paleozoic Rock Units in Arizona

	Northwestern Arizona	Grand Canyon	Central Arizona	Northeastern Arizona	Southeastern Arizona
Permian	Kaibab Limestone; Toroweap Formation; Coconino Sandstone; Hermit Shale; Supai Formation	Kaibab Limestone; Toroweap Formation; Coconino Sandstone; Hermit Shale; Esplanade Sandstone (Supai Group)	Kaibab Ls.; Coconino ss.; Schnebly Hill Fm.; Hermit Shale; Esplanade ss. (Supai Group)	DeChelly Sandstone; Organ Rock Fm.; Cedar Mesa ss. (Cutler Gp.)	Rain Valley Formation; Concha Limestone; Scherrer Formation; Epitaph Dolomite; Colina Limestone; Earp Formation (Naco Group)
Pennsylvanian	Callville Limestone	Wescogame Formation; Manakacha Formation; Watahomigi Formation (Supai Group)	Naco Group	Halgaito Fm.; Honaker Trail Formation; Paradox Formation; Pinkerton Trail Formation; Molas Formation (Hermosa Group)	Horquilla Limestone; Black Prince Limestone (Naco Group)
Mississippian	Redwall Limestone	Redwall Limestone	Redwall Limestone	Redwall Limestone	Paradise Formation; Escabrosa Limestone [Hachita Fm. / Keating Fm.]; Modoc Limestone

This table is intended to show formation names and systemic ages, but not necessarily precise correlations between formations.

Mississippian-Pennsylvanian Disconformity

A time of general uplift and/or withdrawal of the seas occurred throughout Arizona during the late Mississippian or early Pennsylvanian. The result was an erosional surface developed on the predominantly limestone Mississippian age rocks. The common channels, sinkholes, fissures and caves associated with a karst topography buried beneath Pennsylvanian rocks are evidence of this. Numerous boulder and gravel-filled channels on top of the Redwall in western Grand Canyon and on top of the Escabrosa in southeastern Arizona have been documented. Breccia-pipes in the Redwall, formed by collapse of caverns and filling with angular detritus of the overlying Supai and younger rocks, have served as conduits for the passage of ore-bearing (Cu, U) fluids and areas for deposition of ore minerals (see Chapter 15).

Pennsylvanian (325–280 Million Years Ago)

Arizona was generally covered by a shallow sea during the Pennsylvanian with a slow encroachment from southeastern and northwestern Arizona over the relatively high and emergent central portion. The Mississippian-Pennsylvanian boundary is commonly marked by a prominent erosion surface (karst topography) on Mississippian carbonates of the Redwall Limestone, Escabrosa Limestone or Paradise Formation. In southeastern Arizona the basal Pennsylvanian unit is the Black Prince Limestone of Morrowan age, which has a maroon basal shale portion with chert conglomerate lenses derived from the underlying Escabrosa Limestone. The Horquilla Limestone overlies the Black Prince except where the latter is absent it rests on the Escabrosa or Paradise Formations as the basal Pennsylvanian unit. The Earp Formation overlies the Horquilla and spans the Pennsylvanian-Permian boundary.

In central Arizona the basal Pennsylvanian unit is the Naco Formation (Figure 10–9) resting on the eroded Redwall Limestone. In the Grand Canyon, the Watahomigi Formation overlies the erosional surface cut into the Redwall Limestone which is most prominent in western Grand Canyon. One valley of 375 meters (1,238 feet) width and 120 meters (396 feet) depth that was eroded into the Redwall during late Mississippian time was buried by latest Mississippian and early Pennsylvanian sedimentation. The age relations of basal Pennsylvanian rocks across the state provide evidence that the Pennsylvanian seas transgressed across the state from both the southeast and the northwest. Throughout the rest of Pennsylvanian time most of Arizona was covered by shallow seas in which layers of fossiliferous limestones and shales were deposited alternately. Fossils of bryozoans, brachiopods, mollusks, echinoderms and fusulinids are very abundant in most Pennsylvanian rocks across the state (Figure 5–11). In the eastern Grand Canyon area south to the Mogollon Rim the Pennsylvanian strata are predominantly red sandstones and shales that have been assigned to the Supai Group (Figure 10–10). This group has recently been subdivided into four formations, of which the lower three are Pennsylvanian age (Watahomigi, Manakacha and Wescogame Formations). These represent terrestrial to marginal marine deposits that accumulated in the margin of the Pennsylvanian seaway that extended to the south and west. These same stratigraphic units grade westward into a marine limestone called the Callville Limestone in the Virgin and Beaverdam Mountains in northwestern Mohave County. The Pennsylvanian rocks in the northeastern corner of the state consist of marine limestone and evaporites of the Hermosa Group (including the Molas, Pinkerton Trail, Paradox and Honaker Trail Formations) which is gradational laterally into the Supai Group to the west.

Figure 10–9. Naco Formation outcrop near Payson, consisting of fossiliferous marine shale and limestone.

Figure 10–10. Supai Group in Grand Canyon, consisting of predominantly red sandstone, siltstone, shale and minor limestone. The group is subdivided into four formations and transgresses the Pennsylvanian-Permian boundary.

The general tectonic pattern during the Pennsylvanian was continuous subsidence in the northeastern, northwestern and southeastern corners of the state, with the central and eastern portion north of the Mogollon Rim on a tectonically stable region called the Sedona Arch and Defiance Positive (Figure 10–11). The total thickness of Pennsylvanian rocks in Arizona is shown in Figure 10–12.

Permian (280–240 Million Years Ago)

Stable tectonic conditions continued into the Permian much as they were in the Pennsylvanian with marine carbonates and clastics accumulating in southern and western Arizona, but gradually increasing proportions of redbeds, eolian sands and evaporites in the northern and northeastern parts of the state. Fluctuating conditions of transgression and regression are recorded in alternating marine and nonmarine formations across the state.

In the northern part of the state there is a general trend of changing environments during the Permian from the fluvial Esplanade Sandstone and Hermit Shale (Figure 10–13) to the eolian Coconino Sandstone (Figure 10–14), followed by a marine transgression during which the lower

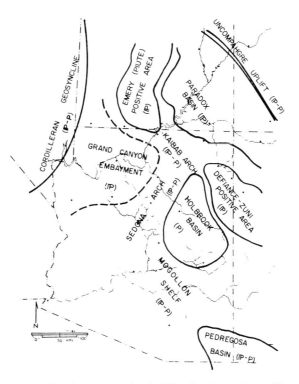

Figure 10–11. Tectonic map showing areas of subsiding basins and stability or uplift during late Paleozoic. Reproduced from Blakey, R. C., 1980. Pennsylvanian and Early Permian paleogeography, Southern Colorado Plateau and vicinity; In Paleozoic Paleogeography of west-central United States Paleogeography Symposium 1, Fouch and Magathan (Eds.) Rocky Mountain Section, Society of Economic Paleontologists and Mineralogists, Denver, Colorado.

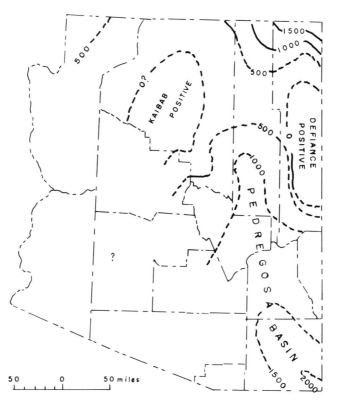

Figure 10-12. Generalized thickness (feet) of Pennsylvanian rocks and location of major tectonic features. Reprinted from Peirce, H. W., 1976, Elements of Paleozoic Tectonics in Arizona: *Arizona Geological Society Digest* 10.

Figure 10-13. Hermit Shale in lower slope with cliff of Coconino Sandstone above.

Figure 10–14. Coconino Sandstone exposure on Hermit Trail, Grand Canyon. **Note prominent eolian crossbedding and contact with overlying marine limestone of Toroweap Formation.**

and middle Toroweap was deposited during a brief regression, after which a transgression occurred resulting in the formation of the Kaibab Limestone (Figure 10–15) and some unknown rocks above the Kaibab which have been eroded away. The evidence for such post-Kaibab Permian strata consists of Permian fusulinids in detrial chert in Triassic age conglomerates resting on the eroded surface of the Kaibab.

Northeastward toward Black Mesa Basin and the Paradox Basin, the Permian rocks change character somewhat and are assigned different formation names, the Cutler Formation and the DeChelly Sandstone. These formations are best exposed in Canyon DeChelly, Monument Valley (Figure 10–16) and the Goosenecks of the San Juan River in Utah. In the northwestern corner of the state the Esplanade Sandstone grades laterally into the Quantoweap Sandstone and the marine Pakoon Limestone.

Permian strata in the southeastern part of the state are predominantly marine including the Earp Formation (interbedded limestone and shale), Colina Limestone (thick bedded, dark gray to black limestone), Epitaph Formation (limestone, mudstone, dolomite and gypsum), Scherrer Formation (predominantly cross-bedded sandstone), Concha Limestone (thick-bedded, cherty limestone) and Rain Valley Formation (thin-bedded, dolomitic and silty limestone). The comparison of these formations with those of northern Arizona is shown in Table 10–2.

The shallow sea in which the middle Permian Rain Valley Formation and Kaibab Limestone were deposited, withdrew from Arizona as the area was uplifted in late Permian time, and erosion was the dominant process at the end of the Paleozoic. An unknown thickness of post-Kaibab—Rain Valley sedimentary rocks was removed, leaving thin deposits of chert and limestone pebble conglomerates in erosional low areas. Some of these conglomerates include Permian fossils younger than those in the Kaibab or Rain Valley formations.

Figure 10-15. Kaibab Limestone near Grand Canyon Village on South Rim of Grand Canyon. Hikers are starting down the Bright Angel Trail.

Figure 10-16. Spider Rock in Canyon DeChelly National Monument. The spire and cliffs in background are eroded into the eolian DeChelly Sandstone.

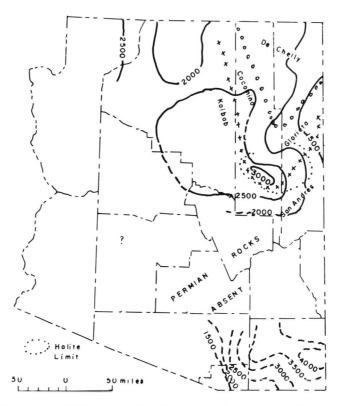

Figure 10-17. Generalized thickness (feet) of Permian rocks in Arizona. Reprinted from Peirce, H. W., 1976, Elements of Paleozoic Tectonics in Arizona: *Arizona Geological Society Digest* 10.

THE MESOZOIC ERA

Triassic (230 to 180 Million Years Ago)

The Triassic Period was a time of general emergence above sea level, following the late Permian withdrawal of the middle Permian shallow seas that covered most of Arizona (Figure 11–1).

No Triassic age sedimentary rocks are known south of the Plateau Province. All of central and southern Arizona was uplifted (forming the Mogollon Highlands) and deeply eroded with the erosional debris spread northward in early Triassic time by sluggish streams to be deposited as mud and sand on the low coastal plain of a shallow sea that extended northward and westward

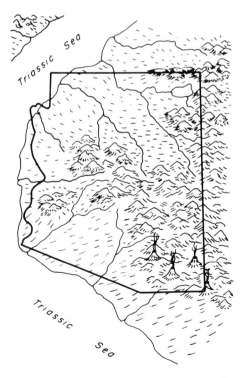

Figure 11–1. Paleogeographic map of Arizona during Triassic time. Reproduced from Wilson, 1962, A résumé of the geology of Arizona; Arizona Bureau of Mines, Bulletin 171.

in Utah and Nevada. This deposit of predominantly red sediments, named the Moenkopi Formation covered an erosional surface on the Kaibab Limestone over most of northern Arizona but onlaps the Coconino and DeChelly Sandstones in northeastern Arizona. The contact between Permian and Triassic strata in northern Arizona represents a hiatus (or missing time) of several tens of millions of years. The Moenkopi Formation is well exposed in Wupatki National Monument, 48 kilometers (30 miles) north of Flagstaff.

By late Triassic time the elevation of the Mogollon Highland had increased and the gradient of the northward flowing streams was increased resulting in the transportation of coarser sediments (sand and gravel) over the coastal plain. These gravels were spread quite uniformly as the Shinarump Conglomerate over a middle Triassic erosional surface cut into the Moenkopi Formation over most of northern Arizona and southern Utah. In northeastern Arizona the Shinarump Conglomerate rests on the DeChelly Sandstone, for example, in Canyon DeChelly.

The Shinarump Conglomerate (Figure 5–6) consists of sand and pebbles of Paleozoic and Precambrian rock units that are known to crop out in central Arizona; therefore, that area must have been the source of the material. The Shinarump sediments are generally coarser to the south and sedimentary structures such as cross-bedding also indicate a southerly source. Commonly included in the Shinarump are petrified logs of conifer trees and associated deposits of uranium minerals (see Chapter 15). The Shinarump Conglomerate is formally classified as the lowermost member of the Chinle Formation in northern Arizona and varies in thickness from 0 meters to over 105 meters (347 feet) depending primarily upon the irregularity of the erosional surface upon which it rests.

Following the deposition of the Shinarump Conglomerate the Mogollon Highlands were evidently reduced in elevation, because subsequently deposited sediments were much finer in grain size. The overlying Petrified Forest member of the Chinle Formation (Figure 11–2) is a thick sequence of gray-green-brown shale, sandstone and volcanic ash, and is locally very fossiliferous, containing petrified trees, amphibians, and reptiles including the earliest North American crocodiles, phytosaurs and dinosaurs. The famous Petrified Forest National Park has been established to protect exposures of exceptionally well preserved examples of petrified trees of the genus *Araucarioxylon* and associated fossils (Figure 11–3). The sediments of the Petrified Forest Member were widespread as volcanic ash falls and stream-transported ash, clay and sand over most of northern Arizona and southern Utah. Most of the fossilized trees were apparently rafted into the area where they were buried, but occasional standing stumps indicate that they grew in the area, and the associated amphibians and reptiles undoubtedly roamed the area during late Triassic time. This rock unit is soft and nonresistant, and typically is weathered and eroded into topographic lows.

The uppermost member of the Chinle Formation, the Owl Rock Member, is gradational into the Petrified Forest Member. It consists predominantly of reddish mudstones, ledge-forming sandstones and fresh water limestones, that were deposited in river flood plains and lakes. Due to its more resistant lithology, it typically forms topographic benches or ridges that stand above the valleys eroded into the Petrified Forest Member (Figure 11–4). The Chinle Formation, consisting in ascending order of the Shinarump Conglomerate, Petrified Forest and Owl Rock Members, is particularly well exposed in the Little Colorado River Valley west and south of Black Mesa where it is eroded into a long, arcuate cuesta with strata dipping gently into the Black Mesa Basin (Figure 11–5).

Figure 11-2. Petrified Forest Member of the Chinle Formation, north of Cameron.

Figure 11-3. Petrified logs of *Araucarioxylon* in the Chinle Formation at Petrified Forest National Park.

Figure 11-4. Owl Rock Member of the Chinle Formation, north of Cameron.

Figure 11-5. Echo Cliffs monocline with exposures of Moenkopi and Chinle Formations in stream valley to right and the Glen Canyon Group in the cliffs. View to south from near Page.

Glen Canyon Group. Some of the most spectacular scenery in northern Arizona is formed by exposures of the predominantly red sandstones and mudstones of the Glen Canyon Group (Figure 11–6), consisting in ascending order of the Wingate Sandstone, Moenave Formation, Kayenta Formation and the Navajo Sandstone. These formations are assigned a late Triassic age, based upon scattered dinosaur remains and tracks. However the position of the Triassic-Jurassic boundary is in dispute with some geologists placing it within the Navajo Sandstone and others placing it lower in the Glen Canyon Group.

Although clearly defined in their respective type sections, the formations in this group grade laterally and vertically into each other, making their recognition difficult for observers who are not intimately familiar with their lithologically variable character.

The Wingate Sandstone is predominantly red, orange and brown cross-bedded sandstone and shale. It is best developed in the Hopi Buttes area, south of Black Mesa, where it is more than 107 meters (350 feet) thick and forms prominent cliffs. It thins in all directions and intertongues westward into the Moenave Formation.

The Moenave Formation is red, orange and brown sandstone and shale, and generally is eroded into valleys and gentle slopes. It is readily observable in its type area near the village of Moenave.

The Kayenta Formation is red-brown to purple sandstone, mudstone and minor limestone, which typically forms prominent cliffs above Moenave slopes (Figure 11–7). It is well exposed in many places on the west, north and east sides of Black Mesa, particularly near Moenave and Tuba City, where tracks of large carnivorous dinosaurs have been found (Figure 5–19). It intertongues vertically and laterally with the Navajo Sandstone. A fine skeleton of the carnivorous dinosaur *Dilophosaurus* was collected from the Kayenta Formation near Moenave, which can be observed in the Navajo Tribal Museum at Window Rock (Figure 11–8).

Figure 11–6. Glen Canyon Group exposed in the Echo Cliffs, southwest of Page.

Figure 11-7. Kayenta Formation exposed in cliffs about 5 miles west of Tuba City.

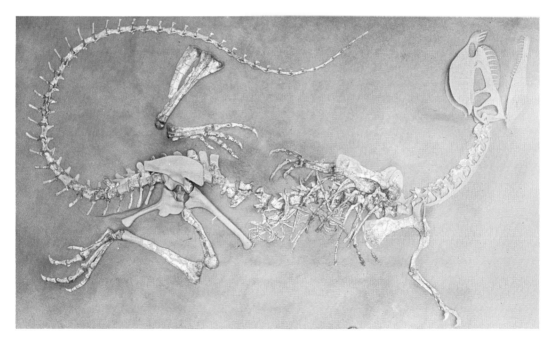

Figure 11-8. *Dilophosaurus* skeleton collected from Kayenta Formation near Moenave. Photo courtesy of Sam Welles, Museum of Paleontology, University of California, Berkeley. Skeleton about eight feet long.

The Navajo Sandstone is buff to orange cross-bedded eolian sandstone, with minor fresh water limestones. It is exposed over broad areas north and west of Black Mesa (Figure 11–5 and 11–6) where it forms the prominent cliffs around Lake Powell and in Zion National Park in southern Utah. Some dinosaur remains and tracks have been found in it. The Navajo Sandstone is the primary aquifer for deep water wells beneath Black Mesa where water for the Black Mesa-Bullhead City coal-slurry pipeline is pumped. Similar rocks of equivalent age and origin (Nugget Sandstone) serve as the primary reservoir rock for several recently discovered oil and gas fields in the Rocky Mountain overthrust belt in Utah and Wyoming.

All the Triassic rock units described in the previous few pages, except the Wingate Sandstone, can be observed along U.S. Highway 89 north between Flagstaff and Tuba City or Page. They are especially well exposed in the Echo Cliffs and Vermillion Cliffs a few miles south and west of Page (Figure 11–5). The Wingate Sandstone and Moenave Formation are formally subdivided into members which will not be discussed here.

Jurassic (180 to 135 Million Years Ago)

San Rafael Group. The San Rafael Group of Jurassic age rests disconformably on the Glen Canyon Group. Like the former, it includes several formations which are lithologically variable and intertongue with each other. In ascending order, the formations are as follows.

The Carmel Formation consists of 30–60 meters (100–200 feet) of gray, red and brown siltstone and sandstone. It rests on an erosional surface on the Navajo Sandstone in northeastern Arizona, and thickens to the north and west where it includes marine limestones in southwestern Utah.

The Entrada Sandstone consists of 60–147 meters (200–485 feet) of orangish to reddish sandstone and siltstone. It is soft and readily weathered and eroded to form valleys along its exposure.

The Summerville Formation is 38–80 meters (125–265 feet) of reddish-orange to brown sandstone and siltstone. It grades laterally from southern Utah into the Cow Springs Sandstone of southwestern Navajo County.

The Bluff Sandstone is a 14–19 meters (47–63 feet) thick, gray, cross-bedded sandstone in northeastern Apache County. It intertongues with the Cow Springs Sandstone and grades laterally into the Summerville Formation.

The Cow Springs Sandstone is 103 meters (342 feet) thick near Kayenta and consists of yellowish gray cross-bedded sandstone. It interfingers with the Summerville Formation and Cow Springs Sandstone. It and the Carmel-Entrada units are eroded into white cliffs on the north side of Coalmine Mesa, 26 kilometers (16 miles) southeast of Tuba City, where they form some of the most beautiful landscapes in Arizona (Figure 11–9).

The Morrison Formation. Along the northern and eastern edges of Black Mesa, the Morrison Formation crops out overlying the San Rafael Group (Figure 11–10). It consists of 150–210 meters (480–690 feet) of varicolored sandstone, mudstone and shale, and wedges out and intertongues to the southwest with the Cow Springs Sandstone. The Morrison is famous for its dinosaur fauna in Colorado and Utah, but has been essentially unexplored in Arizona. The senior author of this book has found large dinosaur bones in the Morrison, but in very inaccessible terrain; therefore they were not collected. It is also the host rock for significant uranium minerals in northern Apache County, from which most of Arizona's production has come (see Chapter 15).

Figure 11-9. Coalmine Canyon, southeast of Tuba City. The Jurassic Carmel-Entrada Formations (undifferentiated) are exposed in lower slope, the Cow Springs Sandstone in the vertical cliff. The dark unit at top of cliff is the Dakota Sandstone of Cretaceous age.

Figure 11-10. Morrison Formation (Jurassic) is exposed in the lower cliff and is overlain by the Cretaceous Dakota Sandstone, Mancos Shale and Toreva Formation. The Toreva Formation contains coal beds, therefore documents a marine regression from the area after the marine Mancos Shale was deposited. Location is Lohali Point, southeastern Black Mesa. Photo by Steve Trimble, Museum of Northern Arizona, Flagstaff.

Triassic and Jurassic rocks of southeastern Arizona consist primarily of plutonic and volcanic rocks, with local occurrences of redbeds. Three major rock units of these ages are recognized south and southeast of Tucson. The older volcanics, consisting of about 3000 meters (9,900 feet) of rhyodacite and andesite, crop out in the Santa Rita, Sierrita and Patagonia Mountains. A stratigraphically higher sequence of redbeds, up to 600 meters (1,980 feet) thick, occurs in the Canelo Hills, Santa Rita, Sierrita, Tucson, Empire and Dragoon Mountains.

The youngest unit consists of 2100 meters (6,930 feet) of silicic flows, tuffs and some clastic rocks in the Canelo Hills, Huachuca, Mustang and Sierrita Mountains. The Triassic-Jurassic age of these rocks has been determined by stratigraphic position between Permian and Cretaceous sediments and a few radiometric dates.

A pronounced erosional unconformity separates Jurassic and Cretaceous rocks throughout Arizona, representing a time of general emergence of the area above sea level.

Cretaceous (135 to 70 Million Years Ago)

Southern Arizona. Lower Cretaceous rocks in southeastern Arizona consist of a thick sequence (3000 meters; 9,900 feet) of conglomerate, sandstone, shale and limestone. These sedimentary rocks were deposited on an erosional surface cut into pre-Cretaceous igneous and sedimentary rocks, in a marine embayment from the Gulf of Mexico called the Sonoran Geosyncline. Relief on the unconformity is considerable, up to 300 meters (990 feet) locally in the Santa Rita and Huachuca Mountains.

During early Cretaceous time a thick sequence (up to 1200 meters; 3,960 feet) of volcanic and sedimentary rocks accumulated in local basins, filling in the irregularities on the Jurassic erosional surface. Overlying these earliest Cretaceous rocks is the widely distributed Bisbee Group and correlative sedimentary rocks (Table 11–1). In the Mule and Huachucha Mountains the Bisbee Group (up to 3000 meters, or 9,900 feet, thick) is made up of four formations including, from the base upwards: The Glance Conglomerate, the Morita Formation, the Mural Limestone and the Cintura Formation. The Glance Conglomerate (or Formation) consists of red, brown and gray sandstone, shale and conglomerate, with intermediate lava flows in the Huachucha Mountains. The Morita and Cintura Formations are repetitious sequences of pinkish gray, arkosic, cross-laminated sandstones that grade upward into massive, grayish red siltstones and mudstones.

The Mural Limestone is a fossiliferous marine unit, in some places forming reefs consisting of rudistid bivalves (extinct oysters with massive calcium carbonate shells) and associated fossils. These reefs form porous masses of limestone of potential reservoir characteristics which make them interesting for oil and gas prospectors.

The Bisbee Group is believed to be a thick deltaic sequence that was deposited in the margin of the Sonoran Geosyncline that extended to the southeast into Mexico. To the northwest in the Tucson area, the marine rocks grade into brackish water and continental arkosic facies.

Upper Cretaceous strata in southeastern Arizona rest on an unconformity cut into the Bisbee Group or older rocks, indicating significant uplift and erosion during middle to late Cretaceous time. In the Santa Rita Mountains, the Fort Crittenden Formation and overlying strata (over 800 meters, or 2,640 feet, thick) contains fossils of fresh water mollusks, fish, turtles and dinosaurs, indicating nonmarine deposition during latest Cretaceous time following the marine deposition of the Mural Limestone.

TABLE 11-1. Mesozoic Rock Units in Arizona

		Virgin Mountains	Black Mesa Basin		Southeastern Arizona	
Creta-ceous	Upper	Cottonwood Wash Formation

Jacobs Ranch Formation | | | Andesite and Rhyolite | |
			Mesaverde Group	Yale Point Sandstone	Silverbell and Salero Fms.	
				Wepo Formation	Amole Arkose	
				Toreva Formation	Pinkard Formation	
			Mancos Shale		Fort Crittenden Formation	
			Dakota Sandstone			
	Lower				Bisbee Group	Cintura Fm.
						Mural Ls.
						Morita Fm.
						Glance Cgl.
Jurassic		Navajo Sandstone	San Raphael Group	Morrison Formation	Undifferentiated Volcanic and Clastic Units	
				Bluff Sandstone		
				Summerville Formation	Cow Springs Sandstone	
				Todilto Limestone		
				Entrada Sandstone		
				Carmel Formation		
			Glen Canyon Group	Navajo Sandstone		
				Kayenta Formation		
Triassic		Chinle Formation

Moenkopi Formation | | Moenave Formation | Rudolfo Fm. and Recreation Redbeds | |
				Wingate Sandstone		
				Chinle Formation	Ox Frame and Mt. Wrightson Fms.	
				Moenkopi Formation		

This table is intended to show formation names and systemic ages, but not necessarily precise correlations between formations.

The youngest of the major units of Mesozoic rocks in southeastern Arizona consists of up to 1350 meters (4,455 feet) of a variety of volcanic rocks that were deposited on an erosional surface cut into rocks ranging in age from late Cretaceous to older rocks.

In summary, the Cretaceous was a time of extensive crustal disturbance, erosion, nonmarine sedimentation and volcanism in southeastern Arizona, with a brief incursion of shallow marine water from the Sonoran Geosyncline to the south. The period ended with an episode of volcanism, and has since remained above sea level. Most of the copper deposits in southern Arizona (see Chapter 15) were formed during this time with the emplacement of plutons of granodiorite porphyry.

The Cretaceous System is poorly represented in western Arizona. Laramide (Cretaceous-Tertiary) age intrusive rocks occur in several of the mountains ranges of Mohave, Yuma and Maricopa counties (for example, Hualapai, Harcuvar and Harquahala Mountains). The ore deposits in many of the famous mining districts of southern and western Arizona were formed with the emplacement of these intrusive rocks (see Chapter 15 for further discussion on mining in Arizona).

Sedimentary rocks of Cretaceous (?) age crop out in Yuma County in the Plomosa, New Water and Dome Rock Mountains. These consist of several thousand feet of redbeds, including sandstone, shale and conglomerate, all of terrestrial origin.

Northern Arizona. Cretaceous rocks of northern Arizona are entirely sedimentary, and are found primarily in the Black Mesa Basin, with small remnants as far south as Show Low. These are all Upper Cretaceous in age, resting on an erosional unconformity cut into pre-Cretaceous rocks, which indicates the area was above sea level during early Cretaceous time when the Bisbee Group was accumulating in the Sonoran Geosyncline in southeastern Arizona. The sequence of formations in the Black Mesa Basin shows a marine transgression (Figure 5–23) followed by a regression in the area during late Cretaceous time. These formations are discussed in the following pages.

Dakota Sandstone—The Dakota Sandstone rests on the erosion surface mentioned above, onlapping successively older formations in a southerly direction until, near Show Low, it rests directly on Permian strata. It crops out around the periphery of Black Mesa at the base of the Cretaceous section where it is divided into three lithologic units, a basal sandstone, a middle carbonaceous unit, and an upper sandstone. The three units are variable in thickness with the carbonaceous unit (containing coal) achieving its maximum thickness (36 meters; 120 feet) in the southwestern part of Black Mesa where it has been mined for local use for many years (Figure 11–11). Well preserved impressions of flowering plants have been collected from several localities. The Dakota Sandstone serves as a reservoir for ground water in some parts of Black Mesa and adjacent areas.

Mancos Shale—The Mancos Shale, exposed around the periphery of Black Mesa, is marine in origin and contains a varied assemblage of fossils including ammonites, bivalves, gastropods, corals, foraminiferans, sharks and plesiosaurs (Figure 5–12). It consists of dark gray to brown silt, clay, minor limestone and fine sand, with several beds of bentonitic clay derived from volcanic ash falls. Its topographic expression is generally a steep slope, commonly deeply weathered and covered by talus from the overlying Mesaverde Group (Figure 11–10 and 11–12). The lower contact with the Dakota Sandstone is gradational and is conventionally picked at the highest sandstone. The upper boundary is gradational with the overlying Toreva Formation and is usually placed at the lowest sandstone of that overlying unit. The Mancos Shale is about 670 feet thick at Rough Rock at the northeastern extremity of Black Mesa, and it pinches out near Show Low.

Figuro 11-11. Dakota Sandstone carbonaceous unit, near Kayenta.

Mesaverde Group—This rock unit consists of three formations containing sandstone, shale and coal in varying proportions. They are from bottom to top: the Toreva and Wepo Formations, and Yale Point Sandstone. The Toreva Formation typically consists of a lower sandstone member, a middle carbonaceous shale member and an upper sandstone member. The total thickness of all three members is about 300 feet. The middle member contains coal in varying amounts. The lower sandstone of the Toreva forms vertical cliffs above the Mancos Shale and huge angular blocks, called Toreva blocks, tend to break off due to undercutting by erosion of the Mancos Shale and slide down the Mancos slope. The upper sandstone member forms the top of Second Mesa located on the southeastern margin of Black Mesa.

The Wepo Formation is a thick sequence of interbedded mudstone, siltstone, sandstone and coal, which varies in thickness 100 to 250 meters (330–825 feet). It is exposed across much of Black Mesa and contains coal beds that are being mined commercially near the northern edge of the mesa (see Chapter 15). The Wepo Formation intertongues with the underlying Toreva Sandstone and the overlying Yale Point Sandstone (Figure 11–13).

The Yale Point Sandstone occurs only along the northeastern margin of Black Mesa, where it forms the cliff-forming cap rock of the mesa (Flgure 11–13). It is about 65 meters thick and is exposed in only a small area on top of Black Mesa.

Figure 11-12. Toreva Formation in cliff at top of Mancos Shale slope, near Kayenta.

Figure 11-13. Wepo Formation and Yale Point Sandstone in upper part of exposure at Yale Point, Black Mesa.

CHAPTER 12

MESOZOIC TO CENOZOIC TECTONICS
AND CENOZOIC ROCKS

The first flourishes of Mesozoic tectonic activity in Arizona began during the Triassic by which time subduction was underway at the continental margin somewhere to the southwest. A magmatic belt was established across the southern part of the state which trended in a northwest-southeast direction. Isolated Triassic plutons and volcanic outcroppings of intermediate to silicic composition are sparsely scattered throughout the Basin and Range Province. Because of the intensity of succeeding events much of the record of this early activity has been obliterated. Existing data indicate that the first phase of magmatism had ended by the end of the Jurassic.

During Triassic and Jurassic time the southern and central part of the state was topographically high and acted as a source area for the sediments that were spread over the Colorado Plateau region. During the Cretaceous the highlands were worn down enough so that coarse sediments began being deposited within local depressions and in the Sonoran Geosyncline. This was a time of relative quiescence in Arizona compared to the thrusting associated with the Sevier Orogeny along the cratonic boundary in southern Nevada and western Utah. To the west was active intrusion of the Peninsular Range Batholith in southern and Baja California.

The Laramide Orogeny began in Arizona in the latest Cretaceous, 75 or 80 million years ago, and waned by about 50 million years ago in the Eocene. During this interval the entire state was affected by tectonism, severely so in the southern part. Initially the southern part of the state was uplifted, with sediments accumulating in basins on the Colorado Plateau between other uplifts forming in that region.

The next event associated with the Laramide Orogeny in the southern part of Arizona was the eruption of intermediate to silicic volcanic rocks. The best exposures of these rocks occur in the southeastern portion of the state, for example the Glory Hole volcanic sequence in the vicinity of Copper Creek in the Galiuro Mountains. Whether or not this volcanism extended westward throughout the Basin and Range is not apparent for little evidence has been recognized from the record. With recent radiometric dating a number of the rocks mapped as Cretaceous volcanics on the State Geological Map have been shown to be of younger, post-Laramide age.

This volcanism was rapidly followed by intense compression directed in a northeast-southwest direction. The subduction which had been affecting the southern part of Arizona throughout the Mesozoic had finally softened the crust enough that it collapsed in its higher levels, causing folding and intense thrust faulting. Rocks were transported perhaps 100 kilometers (63 miles) from the southwest to the northeast over younger units. Once again the best evidence is in the southeastern part of the state, and even there the Laramide faults are obscured by later formation of the Basin and Range and, in many cases, by successive movements on the faults themselves. Portions of these Laramide thrust faults probably occur in such ranges as the Chiricahua, Swisshelm, southern Dragoon, Huachuca, Santa Rita and Rincon Mountains.

Finally the southern part of Arizona was invaded by a number of large plutons ranging from granite to diorite in composition. Examples include central portions of the Santa Catalina, Santa Rita, and Sierrita Mountains. The Kitt Peak Observatory is built on an impressive massif of Laramide granodiorite (Figure 12–1). The face exposed in the Coyote Mountains to the west of the approach road is composed of gneiss which was metamorphosed in association with the intrusion, and in some layers melted, producing the distinct banded appearance.

The Laramide intrusive episode was of fundamental importance to the state of Arizona for it was accompanied by fluids which caused the mineralization of most of Arizona's copper deposits. In some cases the fluids were localized along folds or fault intersections produced during the preceding deformation. Elsewhere the solutions deposited their copper ores in the upper levels of the intrusions. These ores are known as *porphyry copper deposits*. The name derives from the fact that the first such discoveries were in granites or granodiorites which were porphyritic. The name stuck even though some of the deposits have been found not to have phenocrysts in them. Mining districts in porphyry copper deposits include Ajo, Bagdad, Esperanza, Miami, Morenci, Ray, Safford, San Manuel and Silver Bell (Figure 12–2). Discussion of the nature of mineralization and the importance of mining to Arizona are taken up separately in Chapter 15.

Figure 12–1. Kitt Peak, near Tucson. An intrusive body of granodiorite that was emplaced during the Laramide Orogeny.

Figure 12–2. Inspiration Mine—an example of an open pit copper mine, near Miami.

The Laramide Orogeny affected the northern part of Arizona in a less severe way than it did in the south, but the results are what gives the Colorado Plateau its unique setting of deep canyons, long cliffs, and broad expanses. No plutonism or volcanism from this time is recognized in northern Arizona. Instead the area was uplifted to various levels along a series of nearly vertical faults, some with a high-angle reverse motion, indicating that the area was subjected to compression along with differential uplift.

The geological setting of the Colorado Plateau is one of a Precambrian crystalline basement overlain by nearly flat-lying Paleozoic and Mesozoic sedimentary rocks. The faulting arose in the brittle Precambrian rocks, with displacements of up to 1500 meters (5000 feet) at places. As was stated in the chapter on igneous rocks, temperatures rise at depth and rock tends to become more plastic. On the Plateau the sedimentary rocks overlying the Precambrian tended to be *less* brittle than the rocks beneath them. This was not because they were hotter, but because they had not been hardened by recrystallization like the Precambrian rocks below.

The consequence was that the Paleozoic and Mesozoic sedimentary rocks were draped over the fault blocks at depth forming *monoclines,* or one-sided folds. The Colorado Plateau contains probably the best and most numerous examples of monoclines in the world. They were recognized by the pioneering geologists in the region including John Wesley Powell who pictured them in an early report (see Figure 12–3).

Many of the roads on the Plateau follow close to these major linear features. Coconino Point west of Cameron spectacularly displays beds of the Kaibab Limestone and Coconino Sandstone down-draped to the east along a monocline with segments trending both north and west. Highway

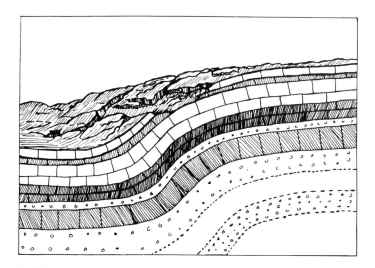

Figure 12–3. John Wesley Powell's (1875) drawing of a typical Colorado Plateau monocline. From *Exploration of the Colorado River of the west and its tributaries:* Washington, D.C., Government Printing Office.

64 between Cameron and Desert View at the Grand Canyon climbs over this monocline to the north where it splits and becomes the Grandview and East Kaibab monoclines. The latter is the eastern boundary of the Kaibab Uplift through which the Colorado River has eroded the Grand Canyon. It is also ascended by Highway 89A, going west toward Jacob Lake.

Another well displayed monocline occurs along the eastern margin of the Defiance Uplift along Routes 12 and 7 north from Lupton along the New Mexico border. The fold at Hunter's Point is especially prominent.

In some cases the uplifted side of the monocline has been eroded away so that all that is seen is a ridge of steeply inclined sedimentary rock jutting out of the ground. This can be seen in the Echo Cliffs monocline along Route 89. (Figures 11–5, 11–6) Fine, flat-iron erosion occurs at Tsegi on Route 160 where the Organ Rock monocline divides northward from the Comb Ridge monocline, an eroded segment that extends northeast past Kayenta and on into southern Utah (Figure 12–4).

The traces of monoclines across the Colorado Plateau probably reflect weaknesses in the underlying Precambrian rocks, established at the time of their formation, along which faulting was concentrated during the Laramide displacements. Although the block uplifts and accompanying development of the monoclines is thought to be primarily a manifestation of the Laramide Orogeny, evidence for dating these events is meager and it is possible that differential uplift on the Plateau continued on into the middle or even late Tertiary.

Following the paroxysm of the Laramide Orogeny, which abated about 50 million years ago, a period of quiescence set in that lasted throughout the Eocene. The heat that had been building up in the crust throughout the Mesozoic due to subduction was transported upward with the rising Laramide magmas. Following this release the crust again stabilized. But subduction continued and heat built up leading to further tectonic activity.

Figure 12–4. Organ Rock Monocline, a Laramide age structure northwest of Kayenta.

During the quiet period the southern part of the state, which had been more strongly affected and uplifted, was eroded to a terrain of low relief with sediments being shed off toward the north, carried to lakes in Utah and Colorado. Deposits of gravel along the southern margin of the Colorado Plateau, the so-called "rim gravels," accumulated during the early Tertiary. The clasts indicate a source to the south because certain rock types found in the gravels are not exposed to the north.

Deposition within the southern part of the state on the Eocene erosion surface was mainly fanglomerate and red fluvial sandstones. Since deposition was in local basins, the rocks vary considerably from place to place and have been given different names accordingly. In the Tucson area this unit is called the Pantano Formation and occurs on the northeast side of town adjacent to the Santa Catalina Mountains, as well as between the Rincon and Empire Mountains. The Whitetail Conglomerate is found in the area south of Superior. In the Phoenix area the Camel's Head Formation crops out on the head of Camelback Mountain and in the buttes of Papago Park (Figure 12–5, 12–6). These and the rocks on Mt. McDowell, or "Red Mountain," east of Phoenix are probably related. In places these Eocene-Miocene fluvial deposits merge into lacustrine deposits of siltstone or limestone, indicating the existence of lakes in some basins at the time. A good example of this lateral gradation occurs in the Artillery Formation in the Rawhide and Artillery Mountains east of Parker (Figure 12–7).

TABLE 12–1. Tertiary Rock Units in Arizona

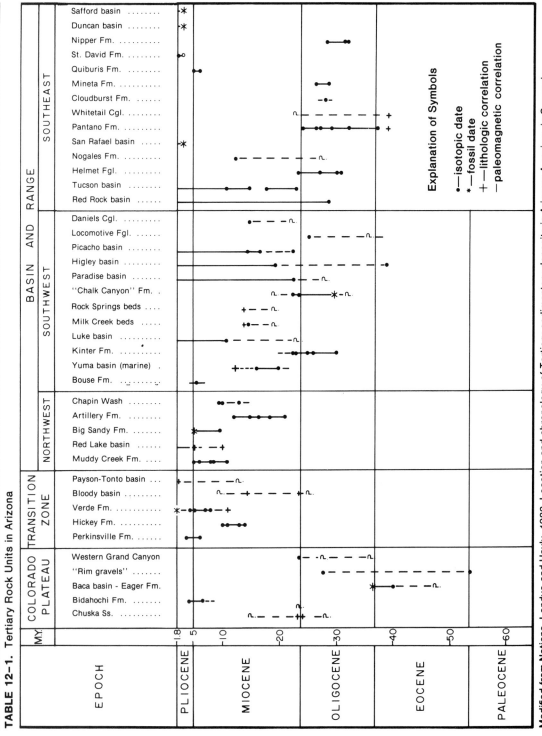

Modified from Nations, Landye and Hevly, 1982. Location and chronology of Tertiary sedimentary deposits in Arizona: A review, in Cenozoic Nonmarine Deposits of California and Arizona, SEPM, Pacific Section.

Figure 12–5. Camelback Mountain, Phoenix. Tertiary sedimentary rocks (Camel's Head Formation) unconformably overlying Precambrian granite.

Figure 12–6. Tertiary sedimentary rocks, Papago Park, Phoenix. The pock-marked differential weathering is called tifoni.

Figure 12–7. The Artillery Formation in western Arizona near Alamo Lake is an example of Tertiary lacustrine and fluvial basin-fill sediments.

In the late Oligocene a period of intense tectonism began which left a profound mark on the southern part of Arizona. This episode, which has been called the "mid-Tertiary Orogeny," was a period of extensive crustal melting, widespread and varied volcanism, formation of metamorphic core complexes, low-angle gravity-induced faulting, and deposition of thick sequences of clastic sediments in local basins.

Voluminous volcanism lasted from about 30 to 20 million years ago. The peak of activity appears to have shifted in a westward direction, from 32 million years ago in southwestern New Mexico, to 26 million years ago in the eastern portion of the Arizona Basin and Range, to 21 million years ago in the Sonoran Desert region. Explosive outpourings of ash and the subsequent formations of caldera complexes are typical of this episode of volcanism (see Chapter 13). Important volcanic centers occur in the Chiricahua, Galiuro and Superstition Mountains; others are scattered throughout the southern part of the state. The composition of these volcanic rocks ranges from rhyolite to andesite and basaltic andesite, but the bulk of the material is toward the silicic end of the spectrum.

The volcanism was only the surface manifestation of magma generation deep in the crust. Magmas that did not make it to the surface remain as plutons emplaced at various depths. Sometimes intrusions occurred into the basal regions of the large volcanic sheets. Sometimes they were associated with the development of metamorphic core complexes.

The *metamorphic core complexes* are composed of intrusive or metamorphic rocks which upwelled and at the same time were markedly extended in a northeast-southwest direction. Ranges built of these features lie in a northwest trending zone from Parker to Tucson, and include the Buckskin, Rawhide, Harcuvar, Harquahala, and White Tank Mountains, the Buckeye Hills, South Mountain, and the Picacho, Tortolita, and Santa Catalina-Rincon Mountains. Other ranges in the southeastern and southcentral part of the state include the Comobabi and Pinaleno Mountains. The northeast trend of the ranges between Phoenix and Parker is in strong contrast with the general north to northwest trend of most of the ranges in the Basin and Range of southern Arizona.

Recognition of the characteristics and extent of core complexes did not come till the late 1970's, demonstrating that discoveries of regional importance can still be made through careful field investigations. Certainly much work is left to be done on the core complexes, for controversy still remains on various aspects of timing and the significance of major features. Previously many of the core complexes had been mapped as Precambrian, and in fact many of them were Precambrian schists and gneisses prior to the new thermal event in the mid-Tertiary, which produced the unique combination of structures in them. Once again radiometric dating has combined with field mapping to elucidate the geological events.

In the core or deepest structural level of the core complexes, are high-grade amphibolitic gneisses and undeformed to highly foliated intrusive rocks. The gneisses often have been plastically deformed with the development of flow folds. These rocks generally originated during Precambrian time or the late Cretaceous-early Tertiary Laramide event.

As one moves upward in these rocks there becomes increasingly prevalent mylonitic foliation, which overprints any older structures. A *mylonite* is a rock in which there has been differential movement, causing the individual mineral grains to be broken or crushed against one another. However, the process takes place at high enough temperature and pressure so the rock remains coherent during the deformation. At low levels the crushing is concentrated in spaced foliation planes with undeformed portions of rock between, but the frequency of these planes increases upward until eventually the entire rock is a mylonite and nothing of the character of the original material remains.

The foliation is gently dipping and produces broad, up-arched areas. Within the foliation planes of the mylonite is a distinct lineation, produced by the elongation of mineral grains invariably oriented in a northeast-southwest direction. This gently-dipping mylonitic foliation and northeast lineation are the most distinctive characteristic features of the Arizona core complexes.

The up-arched foliation can be seen at considerable distance, for instance at the western end of the Santa Catalina Mountains north of Tucson (see Figure 12–8), or in the Picacho Mountains east of Interstate 10. Usually the ranges underlain by the core complexes have gently arched profiles reflecting this foliation, and seldom do they have sharp peaks. South Mountain is a good example, in contrast to the jagged peaks elsewhere in the Phoenix Basin like those in the Sierra Estrella, Phoenix or McDowell Mountains.

In southeastern Arizona and in the Rawhide Mountains in the western part of the state, where late Precambrian and Paleozoic rocks overlie the older Precambrian unconformably, the mylonite zone is often concentrated above the unconformity, with the sedimentary rocks showing various degrees of metamorphism. Elsewhere these strata are not present and the mylonites are developed in the basement rocks.

Figure 12–8. Arched gneiss of the metamorphic core of the Santa Catalina Mountains, Tucson.

In general the mylonite zone becomes fractured upward, eventually to the point that the rock is a breccia which obscures the mylonitic texture. The breccia is often surrounded by the low-grade metamorphic mineral, chlorite. The top of the breccia is so extensively fractured that it is called a microbreccia. The upper surface of this layer is hard and smooth, and acted as a dislocation surface on which overlying rocks glided off the rising core. Grooves and scrape marks called *slickensides* demonstrate that movement occurred along this surface.

Above the dislocation or detachment surface the rocks are unmetamorphosed and tilted along numerous normal faults. The normal faults are relatively steep at high levels, but flatten downward and merge with the detachment horizon. This geometry demands that the cover rocks be rotated during the faulting. The cover rocks in the southeastern core complexes of Arizona are often Paleozoic sedimentary rocks, while elsewhere they are clastic sedimentary rocks that accumulated during the Eocene-Oligocene erosional episode. However, in places the rocks in the cover are as young as Miocene, setting a maximum age for the detachment. A few of the dates on the core complexes extend back into the Eocene, but it appears that most of the thermal activity that led to their formation occurred between 20 and 30 million years ago, synchronous with the major mid-Tertiary volcanism. The dislocation, normal faulting, and tilting apparently somewhat followed the development of the mylonites, between 15 and 25 million years ago. Where the cover rocks are Paleozoic strata, they sometimes have glided off the structural highs as coherent masses cut by few normal faults. When this happened, cascading or attenuated folds were produced. Good examples can be seen beyond the picnic ground at Colossal Cave at the southwestern end of the Rincon Mountains (Figure 12–9) and in the Little Harquahala Mountains (Figure 12–10).

Not all geologists agree with the interpretation of the characteristics outlined above; however there is general accord on the following points. First, the core complexes represent localized thermal upwellings, whether they be strictly plutonic intrusions, or updoming of metamorphic rocks

Figure 12-9. Cascade folds in Paleozoic limestone, Collosal Cave area, Rincon Mountains near Tucson.

Figure 12-10. Metamorphosed and structurally deformed Paleozoic rocks in the Little Harquahala Mountains that have glided off of the rising core complex of the Harquahala Mountains in the background.

themselves sometimes partially melted. Secondly, a great deal of extension was involved, stretching the complexes in a northeast-southwest direction, producing the mylonitic fabric and strong lineation that is seen. Heat seems to have been concentrated at this level, causing metamorphism and promoting the development of the mylonites. Gravitational instability produced by the uparching of the complexes caused the cover rocks to slide off the higher terrain. Movement was concentrated along the foliation planes in the mylonite, perhaps with some lubrication produced by fluids involved in the metamorphism below the detachment (Figure 12–11).

During the mid-Tertiary Orogeny the landscape evolved from relatively low-relief terrain, eroded during the Eocene-Oligocene tectonic lull, to an area of high relief festering with intrusive bulges and volcanic piles. Gray or brown fanglomerates and other coarse-grained clastic sediments were deposited in local basins between the highlands. This deposition continued into the middle Miocene as the mid-Tertiary Orogeny waned. Formations of this period include the Chapin Wash Formation in the Rawhide and Artillery Mountains and the Daniels Conglomerate in the Ajo area.

The middle Miocene was a period of transition in southern and western Arizona. Rotational faulting and tilting of beds came to an end at somewhat different times in different places. Sediments that remain essentially flat-lying today were deposited over locally developed erosional unconformities. Silicic volcanism ceased and was replaced by basaltic volcanism. But most importantly the area began to be broken up into deep basins and high ranges along steeply dipping normal faults without rotational components.

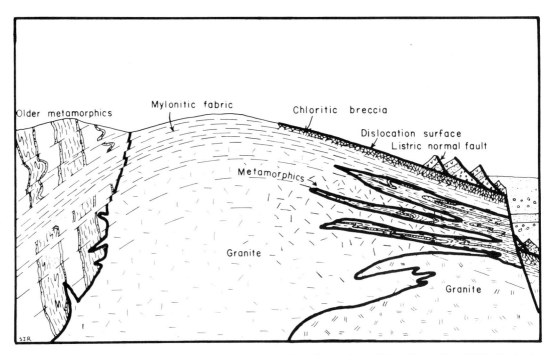

Figure 12–11. Diagram of a metamorphic core complex. Reproduced from Reynolds 1980, Geologic framework of west-central Arizona: *Arizona Geological Society Digest* 12.

The amount of vertical displacement was tremendous, often between 1800 and 3000 meters (5,940 and 9,900 feet) in individual basins. Vertical displacement of equivalent rocks between the edge of the Colorado Plateau and the bottom of the basin south of Mesa is more than 6000 meters (19,800 feet)! Based on relatively uniform altitudes on top of some ranges, it is thought that the Basin and Range province formed more by collapse of the basins than uplift of the ranges. The faulting and subsidence probably began sometime after 15 million years ago and had largely ended by 8 million years ago.

Many of the basins had closed drainage systems during the early stages of subsidence, and various evaporite deposits accumulated in the resulting saline lakes. Halite, gypsum and anhydrite are common minerals. One such deposit is the Luke Salt Body located beneath alluvial cover in the basin west of Phoenix. The Arizona Salt Company is active in recovering salt for the commercial market.

As the uplift came to an end, the basins were gradually filled and through-going river systems were established. The range fronts retreated and broad pediments formed. In most cases the boundary faults between the ranges and basins were buried by alluvium deposited over the distal portions of the pediment surface.

Basin and Range faulting contrasts with the faulting during the mid-Tertiary Orogeny in that it is steeper and has no rotational component. One interpretation is that during the earlier event the deeper portions of the crust were behaving in a plastic manner due to the thermal activity associated with the orogeny, with only the upper portions behaving brittlely and faulting while the deeper rock stretched. As the mid-Tertiary Orogeny abated the crust cooled and became brittle to greater depths, so that when the Basin and Range extension began the resulting faults penetrated to much greater depths with near vertical dips.

This interpretation is supported by the fact that the volcanism associated with the Basin and Range activity is basaltic in composition, derived from sources in the upper mantle, which could have been tapped by the deep faults. This volcanism appears to have ended about 8 million years ago, followed by about 4 million years with little activity. Then in the last four million years renewed activity created several large basalt fields including the San Francisco and White Mountain volcanic fields on the southern margin of the Colorado Plateau, the Sentinal Plain-Arlington field near Gila Bend, the San Carlos field famous for its gem quality peridots and the Pinacate field located at the head of the Gulf of California, encroaching on Arizona in the vicinity of Lukeville.

In the Yuma basin and lower Colorado River area the ocean encroached during the late Miocene, withdrew and then returned in the Pliocene depositing marine siltstones and sandstones. The Pliocene sediments are called the Bouse Formation, which begins with a basal limestone and then upward becomes siltstone, sandstone and tuff.

Cenozoic Basins

The down faulted grabens that began to subside during the Basin and Range Disturbance have been the sites of accumulation of extensive deposits of clastic and volcanic sediments that were eroded from the adjacent elevated ranges. Those deposits have been mapped, described and given formation names which are shown in Table 12–1 (page 144). The ages of the formations are also shown in relative and absolute terms along the left margin of Table 12–1. Quaternary age deposits are not shown in the table, but consist primarily of unconsolidated sand and gravel and lava flows in the San Francisco and White Mountain volcanic fields. The associated volcanic activity along active boundary faults contributed large volumes of lava and ash to the basin fill sequences. There were extended periods of interior drainage, in which thick deposits of lacustrine carbonates and evaporites accumulated. Most of these basins are buried by Quaternary alluvium and the Tertiary rocks are known only from wells that have been drilled into them for water, oil or mineral exploration.

Several Cenozoic basins in southern and western Arizona have been shown recently to contain several thousand feet of clastic, volcanic, calcareous and evaporite sediments, for example Red Lake, Luke, Higley, Picacho and Tucson Basins. Several other Cenozoic basins were formed by faulting and erosion along the southern and western margins of the Colorado Plateau but apparently did not subside to depths as great as those mentioned above. These basins including the Chino Valley, Verde Basin, Payson Basin, Tonto Basin and San Carlos Basin, apparently were relatively stabilized by early Pleistocene and began to be eroded, resulting in the exposure of the Tertiary basin fill sequences within them.

One of the best known examples of Tertiary basin fill sequences is the Verde Formation in the Verde Basin. It contains the best exposed record of Tertiary history in Arizona with about 533m (1600 feet) of Tertiary age sedimentary and volcanic rocks exposed as a result of erosion by the Verde River and its tributaries. The basin fill sequence, called the Verde Formation, consists of approximately 1000m (3000 feet) of interbedded clastic, volcanic, carbonate and evaporite sediments that accumulated throughout the Miocene and Pliocene Epochs. The basin was formed by Oligocene time as a result of faulting on the Verde Fault and probably other faults, and erosion by the ancestral Verde River. By Miocene time the drainage system was blocked at the southern end of the valley by faulting and volcanism, resulting in a closed drainage system. Sedimentation proceeded within the basin with marginal clastic materials grading laterally to lacustrine limestones toward the center where ponded water accumulated periodically. During especially dry periods, the water became hypersaline or completely evaporated, leaving deposits of gypsum and salt in the lowest portion in the southern end of the valley south of Interstate Highway 17. Evaporites and volcanic sediments continued to be deposited in the southern end of the basin throughout the Miocene while the drainage continued to be blocked by faulting and volcanic eruptions in the Hackberry Mountain and Thirteenmile Rock volcanic center about 12 miles south of Camp Verde.

During Pliocene time the cessation of volcanism and faulting in the area allowed the ponded water to spill over and restore external drainage from the basin. The improved drainage prevented the complete evaporation of water and deposition of evaporite minerals as in the Miocene. Fossil evidence indicates shallow water environments were prevalent during the Pliocene, probably resulting from the rapid accumulation of carbonate and clastic sediments which kept pace with basin subsidence. By the end of the Pliocene, at least 1000m (3000 feet) of lacustrine and fluvial sediments had accumulated in the Verde Basin.

By early Pleistocene time external drainage was so well developed that erosion was the dominant process in the basin, and apparently no more significant ponding occurred. Erosion by the Verde River and its tributaries has continued to the present, resulting in the removal of much of the Verde Formation and the creation of the excellent exposures visible today along the edges of numerous mesas in the Verde Valley. Many fossil localities have been discovered in the Verde Formation containing records of the plants, invertebrates and vertebrate animals that lived in the Verde Valley during the late Tertiary.

PART III

TOPICS OF SPECIAL INTEREST

Many subjects of special interest are encountered when discussing the geology of Arizona. We assume that many of them have already received adequate treatment in previous pages of this book. However a few other topics which, in our experience, are of special interest to people will be discussed in the following pages. These topics include: volcanism in Arizona, Meteor Crater, economic geology and environmental geology.

CHAPTER 13

VOLCANISM IN ARIZONA

Volcanism is the most spectacular of geological processes for it occurs in a time frame observable by humans, and often is accompanied by fire, billowing clouds and thunderous explosions. More than one-third of the state of Arizona is blanketed by volcanic rocks, most of which have been erupted in the past 30 million years. The most recent volcano in the state is Sunset Crater, located northeast of Flagstaff on Route 89, in the San Francisco volcanic field (Figure 13–1). Tree-ring dating by researchers at the University of Arizona indicates that the eruption occurred in 1064 A.D.

Because volcanic activity has continued intermittently in the San Francisco Peaks (Figure 13–2) area for the last 1.8 million years, it would seem unlikely that the volcanic field is yet dead, or that Sunset Crater will be the last volcano to erupt in the state.

Volcanic Processes

A volcanic eruption occurs when magma and gases rise from their place of origin within the crust or upper mantle and pour out or extrude at the earth's surface. If the magma flows as a liquid from the volcanic vent it is known as *lava* or a *lava flow*. If it erupts explosively and is blown into many pieces, small and large, it is known as a *pyroclastic eruption* and the fragments

Figure 13–1. Sunset Crater, a cinder cone of the San Francisco volcanic field.

Figure 13–2. San Francisco Peaks, a stratovolcano. View is from northeast into the Interior Valley, which probably was formed by collapse of the summit of the volcano, and then rounded into U-shape by the action of glaciers during the Pleistocene. Sugarloaf Mountain is the tree-covered dome to the left of the mouth of the Interior Valley.

which are produced are known as *tephra*. The nature of eruptions varies greatly, as do the accumulations of the erupted magma. The two most important parameters that govern the type of volcano which is produced are magma chemistry and gas content.

The composition of magmas varies from silicic to mafic, as do the igneous rocks that form from the cooling of the magmas. As the chemistry of a magma varies, so does the physical property known as *viscosity*. Technically, viscosity is a measure of the resistance of a fluid to flow. Remember that silicic magmas are the ones that are highly viscous and form light-colored rocks, and mafic magmas have relatively low viscosity and form dark-colored rocks. This means that rhyolite lava flows are usually thick and stubby, whereas basalt flows are thinner and cover much broader areas (Figure 13–3). An interesting contrast can be observed near Flagstaff, where the later silicic flows of Mt. Elden failed to flow to the base of the mountain (Figure 13–8), whereas the basaltic flows of Merriam Crater extended for about 10 miles to the Little Colorado River where they filled the river channel, creating a natural dam that has resulted in the formation of Grand Falls (Figure 13–18).

The popular notion of a rushing river of molten lava is realized only near the vent of the volcano or at times in the main arteries of the lava flow. The more common scene is that of a lava flow, encrusted by solid rock, advancing slowly and deliberately across a terrain. Because lava is very hot it cools quickly and solidifies upon exposure to the air. The rock at the surface of a flow serves to insulate its inner portions, which remain liquid as long as more lava is supplied. For this reason most of the motion in a lava flow occurs in its interior, with the front advancing as it is

Figure 13-3. Dark feature in center of photo is a basaltic lava flow that issued from the base of SP Crater and flowed northward.

pushed forward by the lava from within. In addition, viscosity increases as the temperature of a lava decreases, so flows tend to move more slowly as they cool.

As a lava flow of basalt moves out from its source, the surface of the flow changes in character. Close to the source the lava is still very hot, and the cooling crust which forms on it is thin and soft. As the lava moves it tends to wrinkle this surface layer or to break through it producing fresh lobes on the top of the flow. The term used to describe this fluid or plastic-looking texture is *pahoehoe* (pronounced "pa-hoy-hoy), and a pahoehoe flow is one whose surface has this texture.

As the flow advances it continues to lose heat, forming a thick, brittle crust, which begins to break into large slabs. Eventually the flow cools such that the crust is very thick. As this occurs the internal flow of hot molten lava continues to drive the flow forward, causing the crust to break into ragged blocks, which tumble down the front of the flow and sometimes are overridden as the flow advances. This blocky texture is known as *aa* (Figure 13–4). Aa flows are more common than pahoehoe, but both types may grade into one another if enough of a particular flow is observable.

Perhaps the best preserved and certainly the most accessible aa flow in Arizona is the Bonito Flow at Sunset Crater National Monument where a self-guided nature trail points out many features of interest.

As lava moves through a growing flow, a network of conduits is established beneath the crusted surface, which connect the volcanic vent with the front of the flow. If the eruption ceases, but the lava within the flow continues to advance, this plumbing system will drain, and open caves known as *lava tubes* will be formed. In order for lava to continue to flow after its source has stopped erupting, the flow must be moving downhill. Such a situation existed when the older basalts of the San Francisco Peaks volcanic field poured down from the plateau into the Verde Valley. Numerous lava tubes were developed in these flows in the area where Highway I–17 climbs north out of the Valley, and other examples are found northwest of Flagstaff.

Basalt flows are a common feature in the Arizona landscape, particularly along the southern margin of the Colorado Plateau and in the central and western portions of the Basin and Range.

Figure 13-4. Aa surface on the Bonito Flow, Sunset Crater National Monument.

The surface features are usually eroded away, but the flows stand out as black or dark gray layers which blanket an area. Sometimes erosion or a well-placed road cut will reveal columnar jointing that formed in the interior of a cooled and solidified flow (Figure 13-5).

Volcanic Features and Rocks

If basalt erupts repeatedly from a vent eventually a volcanic cone is built up. The conical shape is due to the buildup of lava at the vent area with radial distribution of successive flows. Because basalt has a low viscosity, cones built of basalt lava have a very low profile, usually with slopes of 10° or less. Because this flattened cone resembles the shape of a type of shield (used with swords), this type of volcano is known as a *shield volcano* (Figure 13-6).

Shield volcanoes may be small, but if the lava source is longlived and productive, a shield volcano may be tens of kilometers across. Such a volcano exists in northern Sonora, Mexico, at the head of the Gulf of California. It and its subsidiary cones make up the Pinacate Volcanic Field whose northern fringe laps into Arizona in the area between Lukeville and Yuma. Pinacate has been designated a national park by Mexico and may be reached on Mexican Highway 2.

When magma of andesitic composition erupts, because its viscosity is greater than that of basalt, the cone that is built has a steeper profile than a shield volcano. Such a volcano is known as a *composite cone* or *stratovolcano* (Figure 13-13), because andesites often tend to erupt alternately as lava flows and pyroclastic flows and the volcano is a stratified composite of those rock types. The classic, graceful volcanoes such as Mt. Hood in Oregon and Mt. Fuji in Japan are stratovolcanoes.

Figure 13-5. Columnar jointing in basalt near the head of Oak Creek Canyon.

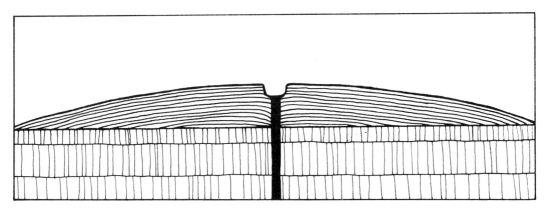

Figure 13-6. Diagram of shield volcano. Diameter up to 100 kilometers.

Arizona has one major stratovolcano, the San Francisco Peaks north of Flagstaff (Figure 13–7). But rather than rising to a single summit the peaks form a ragged boundary to the Inner Basin which breaches the volcano on its northeast side. It is thought that originally the volcano was built to a single peak, but that the summit collapsed into the empty magma chamber below and erosion by streams and glaciers further enlarged the Inner Basin. If one projects a cone from the outer slopes of the presentday San Francisco Peaks, the mountain might have been in excess of 4600 meters (15,000 feet) elevation prior to its collapse.

Rhyolite lavas, being the most viscous, are not able to flow over a terrain for any great distance, and tend to form thick piles. Often the lava does not flow anywhere, but simply oozes out as a sticky mass above the vent. The volcanic form that is built is called a *dome*. In profile it is like an inverted bowl, somewhat wider than it is high. Domes are small features, seldom exceeding 1 kilometer (.6 mile) in diameter. Mt. Elden, near Flagstaff, is an elongate dacite dome (Figure 13–8).

The common way for a dome to grow is by a process of inflation, with lava being added to the interior at the volcanic vent, and the outer portions being severely distended. The outer portions are solidifying as the dome grows so it becomes mantled by a chaotic rubble of its own making. As blocks at the margins of the dome are broken loose they tumble down to form a skirt of lava fragments, which may be overridden as the dome grows. Cracking in the crust of the dome may reach to its molten core, tapping lava which oozes out to the surface (Figure 13–9).

Thus far we have discussed the variety of volcanic forms produced by lava eruptions. For these the magma involved is low in its content of gas. However, it is common for magma to contain an appreciable amount of gases, such as water vapor, carbon dioxide, chlorine, fluorine,

Figure 13-7. San Francisco Mountain from the south, showing its gently sloping, slightly concave sides which are typical of stratovolcanoes or composite cones. Photo by David Best.

Figure 13-8. Mt. Elden, San Francisco volcanic field. Dome built of viscous lava flows. Photo by David Best.

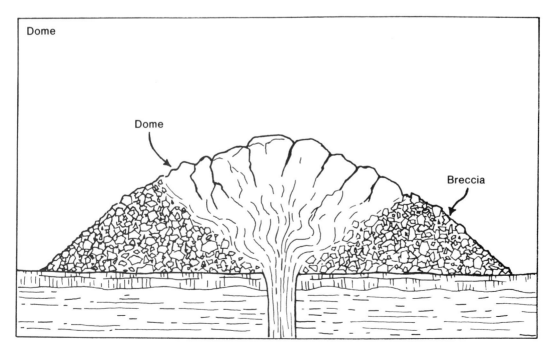

Figure 13-9. Diagram of volcanic dome. Diameter up to 1 kilometer.

hydrogen sulfide and sulfur dioxide. At depth, due to the pressure of the overlying rock, these gases are held in the magma in a compressed state. No void spaces exist. However, during an eruption, as the magma nears the surface and is extruded, the loss of pressure permits the gas to expand as bubbles, or to *vesiculate*. The bubbles become trapped as the lava cools and remain in the volcanic rock that is formed. These spherical or elongate holes are called *vesicles*.

When the gas content is considerable, its expansion causes the magma to be blown into the air as separate fragments and fall to the surface as pyroclastic deposits. In general, the greater the gas content in a magma, the more explosive is the eruption. Though there are inevitable exceptions, gases tend to be more prevalent in silicic magmas than in mafic ones. Therefore, we observe that basaltic (mafic) eruptions are more often of the lava type, and rhyolitic (silicic) eruptions, of the pyroclastic type. As mentioned before, andesites are fickle and often erupt alternately as lava and as tephras (cinder or ash). Also, when basalt does erupt pyroclastically, the explosions are usually less severe than with a more silicic eruption. Here viscosity plays a role as well, for sticky silicic magmas greatly resist the expansion of gas, despite the removal of confining pressure as the magma reaches the surface. The result is that when the gas finally does expand, it does so violently, sometimes hurling debris as high as the stratosphere and forming widespread blankets of ash when it falls to the ground surface. Basaltic pyroclastic eruptions are, in contrast, sputtery and form *cinder cones* and localized cinder blankets (Figure 13–10, 13–11, 13–12).

The volcaniclastic deposits formed by explosive eruptions are called *tephra* which is classified according to size of the fragments. Particles less than 2 millimeters (0.08 inches) in length are called *ash*. This volcanic product is not produced by burning, but is composed of very fine particles of magma that have cooled and solidified as they fell through the air. A consolidated deposit of ash is known as *tuff* (Figure 13–14). Volcanic fragments between 2 and 64 millimeters (0.08 to 2.56 inches) in diameter are termed *lapilli,* and a lapilli tuff is an ash deposit containing some fragments greater than 2 millimeters (0.08 inches) in size. Fragments greater than 64 millimeters (2.56 inches) are known as *bombs* if they are still in the molten state when extruded and are usually rounded or elongate masses of twisted lava. *Blocks* are angular pieces of solid material broken from the volcanic neck or vent area. The special term *cinder* is applied to lapilli and bombs of basaltic composition which are rough-edged, vesicular lumps.

During their extrusion, cinders pile up to form a *cinder cone*. The cone is characteristically smooth-sided, and contains a depression (crater) at its top where the vent was located (Figure 13–12). During a cinder eruption the magma jets out and is blown into fragments which fall around the active vent. Since the cinders are essentially solid when they land they pile up with a characteristic angle on their slopes known as the *angle of repose*. This is the steepest slope that can be maintained by the cinders without their sliding or tumbling down the side of the volcano. Cinder cones have slopes of approximately 30°. Because the cinders are at the steepest angle at which they may be at equilibrium, climbing a cinder cone is literally one step back for every two steps forward.

Cinder cones are a common feature in some of the younger volcanic fields in the state particularly in the San Francisco Peaks and White Mountain areas. The Pinacate Volcanic Field in northern Sonora, Mexico, also has many fine examples. Sunset Crater, located northeast of Flagstaff in the San Francisco Peaks volcanic field, is the state's youngest volcano. It has been designated a National Monument which provides public services such as a self-guided nature trail and excellent camping facilities. Due to destruction of the cone's fragile surface by pedestrian

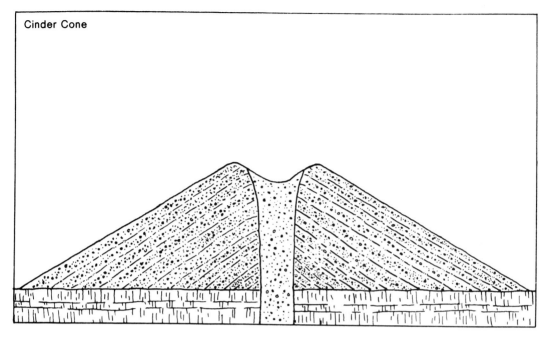

Figure 13-10. Diagram of cinder cone. Diameter up to about 1 kilometer.

Figure 13-11. Red Mountain, located about 30 miles northwest of Flagstaff, is a cinder cone that has been dissected by erosion, exposing its internal structure of stratified cinders.

Figure 13-12. SP Crater, San Francisco volcanic field. Note spatter rampart at lip of crater extending partially into it. Lava flow was erupted through the side of the cinder cone at a late stage of the activity.

traffic, climbing is no longer permitted on the sides of the cone. The name derives from the yellowish-red color of the cinders at the top of the cone, which prompted J. W. Powell, an early explorer, to remark that "it was just like a sunset always on the rim" (Figure 13-1).

Prior to the cinder eruption at Sunset Crater, lava had issued from the same vent area producing an aa flow that covered several square miles. This has been named the Bonito Flow (Figure 13-4). An interesting feature known as a *squeeze up* occurs at places on the flow where the brittle outer shell cracked down to the still molten interior of the flow, allowing the lava to ooze out through the opening.

When erupting magma contains a considerable quantity of gas, the expansion of the gas produces a froth of tiny bubbles, separated from each other by thin walls of glassy magma. Further expansion bursts the bubbles, and the broken pieces, or shards, become ash. This occurs within a few seconds or less at the volcanic vent. The severity of the eruption depends upon the amount of energy released by the expanding gas.

During most pyroclastic eruptions clouds of ash are produced which billow up and away from the eruption. As these clouds become less energetic the ash settles out, blanketing the landscape in a pattern that reflects the wind direction at the time. The deposit which is produced is known as an *ash fall* or an *ash-fall tuff*. Several hundred square miles were covered by ash associated with the eruption of Sunset Crater.

Another important type of ash deposit is an *ash-flow tuff*. During the eruption the ash particles pour over the rim of the vent, and move down the side of the volcano or out over a landscape as a ground-hugging mixture of expanding gases and highly energized particles. This mixture behaves much like a fluid, flowing down valleys and around hills, sometimes lapping over the small ones. Because of the density of particles in the flow, and the presence of hot gases, the flow comes to rest as a thick layer composed of fragments of hot glass. In the interior and lower portions of the flow where the heat is retained longer, and the material is under pressure from the overlying part of the flow, the particles tend to flatten and stick together closing up any pore space. This process is known as *welding* and the rock formed is called *welded tuff*.

An eruption producing an ash flow would also be accompanied by clouds of ash and ash-fall deposits, but it is usually the ash flow that results in the greatest volume of material deposited. Ash flows alternate with lava flows in the building of a composite cone such as the San Francisco Peaks (Figure 13-7, 13-13). However, the most tremendous eruptions on earth must be those which have deposited *ash-flow sheets* (Figure 13-14). Many of the sheets are more than 16 kilometers (10 miles) and some as large as 80 kilometers (50 miles) in diameter. Thicknesses of several hundred feet are not uncommon. No one in historic time has observed such an eruption, so the exact nature of the venting is unknown. However, through careful study geologists have identified a volcanic feature, known as a *caldera,* that is often associated with major ash-flow tuff

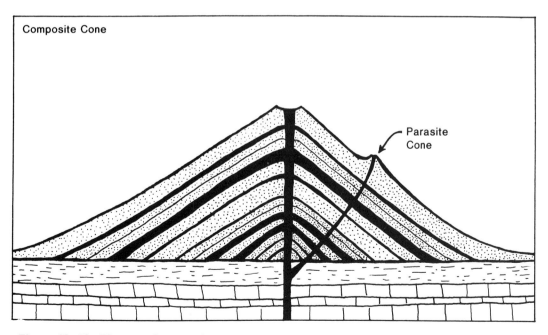

Composite Cone

Parasite
Cone

Figure 13-13. Diagram of composite cone or stratovolcano. Alternating lava flows and pyroclastic flows. Diameter up to 40 kilometers.

Figure 13-14. Ash-flow tuff of the Geronimo Head Formation unconformably overlying Precambrian granite, Usery Mountain Park, north of Mesa.

sheets. Caldera complexes have been mapped in the Superstition and Chiricahua Mountains, and probably exist in some of the lesser known silicic volcanic fields in Arizona.

The general development of a caldera is as follows (Figure 13–15). Silicic magma accumulates in a shallow magma chamber. As more magma is added, pressure is built up and the rock above the magma chamber is arched upward. Eventually it cracks from the internal pressure giving the magma a pathway to the surface. Once the eruption begins it progresses very rapidly so that much of the magma chamber is vented to the surface in a single episode lasting perhaps from several hours to several weeks. During or after the eruption the rock that was capping the magma chamber collapses into the empty chamber along a series of circular or ring faults. The resultant depression is the caldera.

If the collapse occurs after the eruption has ceased then a circular hole is left in the ground. However, often the caldera collapses during the eruption and the ash being produced fills the sinking area with thousands of feet of tuff. If magma continues to be added to the magma chamber after formation of a caldera, the chamber will reinflate and the collapsed roof will be pushed upward, producing a raised area where previously there was a depression. This then is known as a *resurgent caldera*. The main face of the Superstition Mountains (Figure 13–16) northeast of Apache Junction marks the approximate boundary of a resurgent caldera in that volcanic field.

Thus far we have discussed pyroclastic eruptions in which the fragmentation is produced by expanding gases contained within the magma. Another type, the phreatic eruption, bears mentioning. This occurs when magma encounters the groundwater table as it rises toward the surface. The water is turned to steam when the hot magma comes in contact with it and the pressure is released by blowing a hole in the ground above the point of contact with the magma. The depression caused by this type of explosive eruption is called a *maar*. Maars seldom exceed 1 kilometer (0.6 mile) in diameter. The ejecta contains broken fragments of the rock that contained the groundwater, but bits and pieces of the magma are also incorporated in the material that is de-

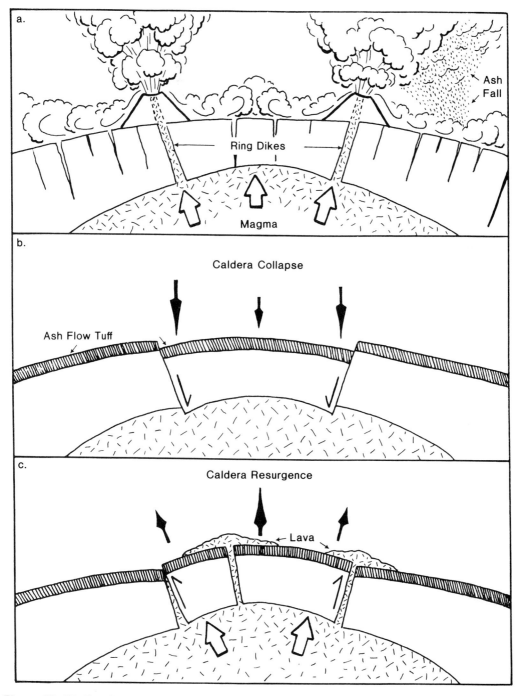

Figure 13-15. Development of a caldera complex. a) eruption of ash-flows form circular fractures above magma chamber. b) collapse of roof of magma chamber during and after eruption forming the circular depression of the caldera. c) refilling of magma chamber from beneath with subsequent uplift or resurgence of the caldera block.

Figure 13-16. Superstition Mountain. The western face marks the approximate western boundary of the resurgent caldera.

posited around the vent. When the explosion occurs, the debris that is thrown out of the maar piles up around the depression producing a *tuff ring*. The explosion occurs as a single detonation or more commonly as a series of explosive pulses, for tuff rings are often stratified, indicating successive waves of particles emanating from the explosive center.

Excellent examples of maars and tuff rings occur in the Pinacate Volcanic Field at Crater Elegante and Cerro Colorado. Another tuff ring is found around Sugarloaf Mountain at the mouth of the Inner Basin of the San Francisco Peaks. A quarrying operation there exposes some of the internal stratification of the tuff ring. Following the initial phreatic eruption, with the lid blown off, so to speak, rhyolitic magma oozed up passively into the maar producing the dome which today is named Sugarloaf Mountain.

Volcanic Fields of Arizona

The San Francisco volcanic field includes the San Francisco Peaks, with Mt. Humphreys the highest point in the state, and Sunset Crater, the youngest volcano, plus about 400 other vents which have been identified in the area. San Francisco Peaks also is one of the two locations in Arizona which were glaciated during the Pleistocene; the White Mountains is the other.

Volcanic activity began about 10 million years ago with extrusion of basalt flows, the so-called Older Basalts that were erupted intermittently until about 4 million years ago, covering a broad area from Flagstaff to the Verde Valley. The long ramp where Interstate Highway 17 climbs north

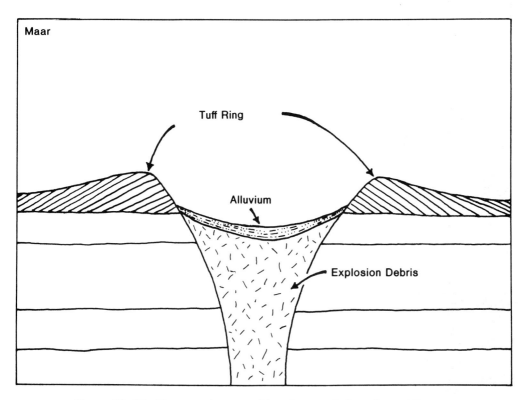

Figure 13-17. Diagram of a maar. Diameter usually less than 1 kilometer.

out of the Verde Valley is underlain by these basalts. Both silicic and basaltic volcanism have occurred intermittently during the past 3 million years.

The San Francisco Peaks are a composite cone of andesite to rhyolite composition. The oldest flow is an andesite which has been dated at 1.8 million years. Dacite, rhyolite, and more andesite were extruded until about 700,000 years ago. The volcano that resulted probably was more than 4600 meters (15,000 feet) in elevation, prior to collapse of its inner portion about a half million years ago. This collapse and subsequent glacial erosion is what formed the Interior Valley of the San Francisco Peaks, which today is an important water source for the city of Flagstaff.

Sugarloaf Mountain, located at the mouth of the Interior Valley, was erupted about 220,000 years ago. As mentioned previously, an initial phreatic eruption produced a tuff ring into which oozed the sticky rhyolite magma that built Sugarloaf Mountain.

Several other volcanic centers in the San Francisco volcanic field are of note. Mt. Elden, which rises north of Flagstaff, is a dacite dome erupted about 550,000 years ago. It is made up of a series of viscous flows which poured out to the south. The patterns of the well developed joints on the mountain's face indicate flow directions. It is thought that some of the lavas may not have reached the surface but inflated the dome from within. On the north side of the mountain the lavas pushed up the overlying Paleozoic sedimentary rocks and tilted them up to 60° in places.

Excellent views of Sunset Crater, the Interior Valley, and all of the eastern San Francisco volcanic field can be seen from the fire tower on the top of O'Leary Peak, a rhyolitic dome that erupted about 230,000 years ago. SP Crater is a nearly perfect cinder cone west of Hank's Trading Post on U.S. 89 (Figure 13–12), 30 miles north of Flagstaff. The crater is rimmed by an excellently developed *spatter rampart*, built of large blobs of lava tossed up as the last sputtery gasps of the pyroclastic phase of eruption. Like Sunset Crater, SP Crater also has a pronounced aa lava flow which issued through the side of the cone and flowed northward for several miles.

Merriam Crater, a cinder cone on the eastern margin of the San Francisco volcanic field, also produced a lava flow in its late stages of activity. This basalt flowed eastward 16 kilometers (10 miles) and eventually poured into the channel of the Little Colorado River, filling it at the entry point and flowing northward in the canyon bottom for a short distance. This event caused a diversion of the river which now spills over the canyon wall downstream of the dam producing Grand Falls, a cascade higher than Niagara. The river is dry most of the year but during times of spring snow melt and summer thunder storms, water cascades over the falls, creating an unusual and imposing sight (Figure 13–18).

Figure 13-18. Grand Falls, formed when lavas from Merriam Crater flowed into the Little Colorado River Gorge, causing a diversion of the river which now spills over the wall of the canyon. Road in the lower left of figure is on basalt which fills the gorge. This basalt also thinly covers the Kaibab Limestone to the left of the meander. The Moenkopi Formation is exposed in the darker, upper right one-fourth of the photo.

Another area in Arizona where lavas have flowed into a canyon is at Vulcan's Throne in the western Grand Canyon south of Toroweap Valley. The basaltic lava flows issued from vents marked by cinder cones on the north rim, cascaded down the near vertical canyon walls, and dammed the Grand Canyon several times, and each time were cut through by the mighty river. The oldest set of flows, dated at 1.2 million years, occurs as remnants on the canyon walls up to 420 meters (1,385 feet) above the present river level. Remnants of the most recent lavas, probably a few tens of thousands of years old, are preserved as cascades on the northern wall of the canyon, where they plunge more than 900 meters (2,970 feet) to the canyon bottom (Figure 13–19, 13–20).

Numerous volcanic necks are preserved in the Hopi Buttes volcanic field north of Holbrook and the Navajo volcanic field in the Four Corners region. Erosion has removed the surrounding sedimentary rocks so that the more resistant volcanic rocks are exposed as buttes that stand above the surrounding terrain. In the Hopi Buttes field the necks are filled with columnar lava whereas the necks in the Navajo Field contain tuff-breccias caused by underground explosions. The most spectacular center in the Navajo volcanic field is Agathla Peak which stands nearly 300 meters (990 feet) above the surrounding topography (Figure 13–21). These volcanic necks contrast with the other buttes in Monument Valley which are erosional remnants of sedimentary rocks which clearly show bedding on their faces.

Figure 13–19. Vulcan's Throne, a basaltic cinder cone on the north rim of the Grand Canyon. Photo by Kenneth Hamblin.

Figure 13-20. Lava cascades at Vulcan's Throne, western Grand Canyon. Photo by Kenneth Hamblin, Brigham Young University.

Figure 13-21. Agathla Peak, a volcanic neck in the Navajo volcanic field north of Kayenta.

The fabled Superstition Mountains to the east of Phoenix are a large caldera complex which developed during the middle Tertiary (Figure 13–22). Remnants of ash-flow sheets occur almost as far south as Florence and to the northeast well past Roosevelt Lake (see Figure 13–14). Three separate caldera structures have been identified in the western portion of the volcanic field. The oldest of the three, the Superstition Caldera, was resurgent, with the central portion arched into a dome as magma refilled the magma chamber beneath. The crest of the dome was down-faulted in a graben structure. The western boundary of the caldera is approximately along the face of Superstition Mountain east of Apache Junction.

The second caldera to form is named the Goldfield Caldera, which contains several blocks of volcanic rock tilted to the northeast. The Tortilla Caldera, which followed, is an arcuate rather than a circular structure with the downwarping occurring along the northeast margin.

Although the eruptive history of the volcanic field is complex, with major dacitic ash-flow sheets, silicic plugs and domes, breccia deposits and basalt flows, there are only two main units, each associated with the formation of one of the older two calderas. The Superstition Tuff was erupted in connection with the Superstition Caldera around 25 million years ago. It blankets

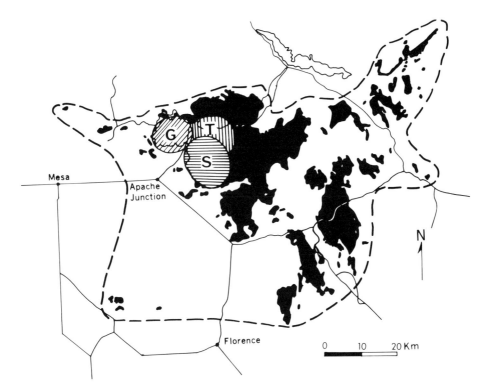

Figure 13–22. Distribution of volcanic rocks associated with the Superstition volcanic field. Superstition caldera (S), Goldfield Caldera (G), and Tortilla Caldera (T) are indicated by the line pattern. Reproduced from: Sheridan, M. F., 1978 The Superstition Caldera Complex *in* Guidebook to the Geology of Central Arizona, Az. Bureau Geol. and Min. Technology, Special Paper No. 2.

much of the southern area with bedded units exposed in the upper face of Superstition Mountain. The younger Geronimo Head Tuff, dated at about 15–16 million years, was erupted from vents in the vicinity of the Goldfield Caldera. This unit crops out over a wide area in the Superstitions, producing characteristic yellow cliffs, such as those seen on the margins of Saguaro Lake, at Usery Pass, and in the canyons in the vicinity of Fish Creek. From the Apache Trail above Apache Lake one can see the yellow beds of the Geronimo Head tuffs exposed along the southern end of the Mazatzal Mountains to the west, and one can imagine the thunderous rush of ash as it blew through the low country at the foot of the older mountains from the Goldfield Caldera to the south.

Although much of the silicic volcanism in Arizona dates from the middle Tertiary, basaltic volcanism has been active until very recently. One must wonder if magmas are not still simmering at places beneath the ground, awaiting the time when they will be released to surge to the surface in yet another volcanic eruption.

CHAPTER 14

METEOR CRATER

As if Arizona were not blessed enough with geological wonders, 56 kilometers (35 miles) east of Flagstaff is an exceptional feature of extraterrestrial origin. Meteor Crater is the world's best preserved meteorite impact structure, produced when a meteorite fell on the Colorado Plateau perhaps 20,000 years ago. The crater is approximately 1.2 kilometers (0.75 mile) in diameter and 180 meters (594 feet) deep with an upturned rim which rises 30–60 meters (99–198 feet) above the surrounding plateau. The bowl-shaped depression is slightly squarish due to a mutually perpendicular joint set which is developed in the bedrock in the area (Figure 14–1).

A thin veneer of Triassic Moenkopi Formation overlies the Permian Kaibab Limestone in the area. These, as well as the upper part of the Coconino Sandstone, are exposed in the walls of the crater. At the time of impact the rocks at the rim were upturned and in places were literally folded back on themselves. Debris ejected from the crater encircles it out to a distance of several miles showing an inverted stratigraphy; that is, the uppermost fragments in the debris blanket are

Figure 14-1. Meteor Crater from the southeast. Note upturned rim and debris strewn from the crater. Photo courtesy of Center for Meteorite Studies, Arizona State University.

Coconino Sandstone, successively overlying fragments of the Kaibab and Moenkopi. The implication of course is that at the time of impact the uppermost layers of rock were hurled out first with deeper layers being brought up in turn. Fragments in the debris range from tiny splinters less than a micron in length up to blocks in excess of 30 meters (99 feet) across. Although the inverted stratigraphy is well preserved, some mixing of the formations does occur, and pieces of the meteorite are found throughout (Figure 14–2).

Drilling has shown that fractured bedrock and fine meteoritic material extend to a depth of 180 meters (600 feet) beneath the floor of the crater. This fractured portion passes upward into a massive, graded unit about 10.5 meters (35 feet) thick which is coarse-grained at the bottom, becoming fine-grained upward. This unit has been interpreted as fallout produced by an explosion at the time of impact which threw debris to a great height into the air, with the larger, heavier pieces settling out first. This unit is overlain by talus and alluvium that have washed down from the walls of the crater since its formation. Lake beds in the center of the crater indicate that standing water was present at times.

The name Meteor Crater is really somewhat of a misnomer, since the term *meteor* is reserved for the features which we know colloquially as "shooting stars," and *"meteorite"* is reserved for a meteor that strikes the earth's surface. Actually these are solid objects, not stars, which upon entering the earth's atmosphere are heated by friction to the point that the outer surface of the meteor melts and luminesces. The objects which become meteors are known as *asteroids*—members of our solar system, revolving around the sun like the planets. They occur throughout the solar system, but are particularly concentrated in a belt between Mars and Jupiter. During their heavenly passage these bodies drift into the earth's gravitational field, at which point they are captured and produce the familiar phenomena of "shooting stars."

Asteroids are thought to be fragments produced at the time of formation of the solar system, and as such are very important for scientists' theories regarding its formation. Unlike the Earth, which has been a dynamic, chemically differentiating body since that time, the asteroids have been inert since their time of creation and therefore preserve a chemical signature of the early cosmic event which produced the sun and its planets. Isotopic dating of numerous meteorites consistently gives an age of 4.6 billion years for their time of formation, and, by implication, the time of formation of the earth and the rest of the solar system.

As the surface of a meteor melts due to friction with the atmosphere, it is left behind producing a bright streak in the sky. The heat that causes the melting is carried away with the molten material, so that when a meteorite hits the ground it is still cool on its interior, contrary to the popular notion that meteorites are too hot to handle when they land. This principle of removal of heat by melting the surface of an object is utilized by spacecraft as they re-enter the earth's atmosphere. The convex, pan-shaped heat shield on the Apollo capsules is the same aerodynamic shape as that found on some meteorites that have fallen through the atmosphere in a fixed orientation without rotation.

Meteorites are classified into three main types, *iron, stony-iron,* and *stony.* Iron meteorites are composed of nickel and iron which combine in two different phases of nickel-iron alloys that sometimes crystallize in an interlocking, geometric pattern known as *Widmanstatten structure* (Figure 14–3). The meteorite that produced Meteor Crater is of this type. Stony-iron meteorites contain approximately equal amounts of nickel-iron alloy and silicate minerals.

The most abundant type is the stony meteorite. These are composed primarily of the silicate minerals olivine, pyroxene, and plagioclase feldspar. They appear much like some earth rocks and

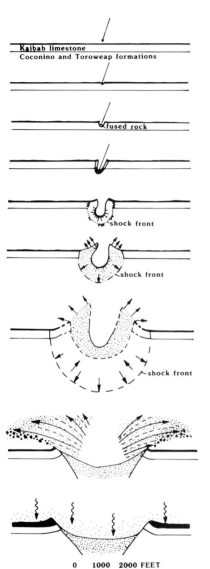

1. **Meteorite** approaches ground at 15 km/sec.

2. **Meteorite** enters ground, compressing and fusing rocks ahead and flattening by compression and by lateral flow. Shock into meteorite reaches back side of meteorite.

3. Rarefaction wave is reflected back through meteorite, and meteorite is decompressed, but still moves at about 5 km/sec into ground. Most of energy has been transferred to compressed fused rock ahead of meteorite.

4. Compressed slug of fused rock and trailing meteorite are deflected laterally along the path of penetration. Meteorite becomes liner of transient cavity.

5. Shock propagates away from cavity, cavity expands, and fused and strongly shocked rock and meteoritic material are shot out in the moving mass behind the shock front.

6. Shell of breccia with mixed fragments and dispersed fused rock and meteoritic material is formed around cavity. Shock is reflected as rarefaction wave from surface of ground and momentum is trapped in material above cavity.

7. Shock and reflected rarefaction reach limit at which beds will be overturned. Material behind rarefaction is thrown out along ballistic trajectories.

8. Fragments thrown out of crater maintain approximate relative positions except for material thrown to great height. Shell of breccia with mixed meteoritic material and fused rock is sheared out along walls of crater; upper part of mixed breccia is ejected.

9. Fragments thrown out along low trajectories land and become stacked in an order inverted from the order in which they were ejected. Mixed breccia along walls of crater slumps back toward center of crater. Fragments thrown to great height shower down to form layer of mixed debris.

Figure 14-2. Diagrammatic sketches showing sequence of events in formation of Meteor Crater. Modified from Shoemaker, E., and others, 1978 Barringer Meteorite Crater, Coconino County, Arizona, *in* Guidebook to the Geology of Central Arizona, Az. Bur. Geol. and Min. Technology, Special Paper No. 2.

Figure 14–3. Polished cut of the Baghdad, Arizona meteorite, an iron meteorite displaying Widmanstatten structure. Note the alteration rim due to heating at the edge of the meteorite. Photo from the collection of the Center for Meteorite Studies, Arizona State University.

therefore are difficult to identify in the field. This group is further subdivided according to the presence or absence of a structure called a *chondrule,* a small, spherical silicate inclusion which suggests melting or recrystallization during the history of the meteorite. Stony meteorites with chondrules are called *chondrites,* and those without, *achondrites.* Certain chondrites, the *carbonaceous chondrites,* contain considerable amounts of water combined with hydrated silicate minerals, as well as complex organic compounds. Although these do not actually indicate that "life" existed in the meteorites, the compounds are similar to certain organic molecules that are the building blocks for biological compounds on earth. The carbonaceous chondrites are considered to be the most primitive of the meteorites, and as such represent the most primordial material left from the event that produced our solar system.

How does one identify a meteorite? Though unlikely, the easiest method would be to see a meteorite hit the ground, go over and pick it up. A meteorite so recovered would be known as a *fall.* A meteorite discovered sometime after it has fallen is known as a *find.* Perhaps the first clue that you might have a meteorite is the weight of the specimen. Iron meteorites are three times heavier than most terrestrial rocks, and stony meteorites about one and one half times. The surface of a meteorite is often smooth and may have various linear or fluted patterns. A common surface texture on iron meteorites are shallow depressions called *"thumbprints"* (Figure 14–4).

Figure 14-4. Iron meteorite from Henbury, Australia, showing "thumbprint" surface texture due to melting from friction with the atmosphere. Cm scale. Photo from the collection of the Center for Meteorite Studies, Arizona State University.

The Social History of Meteor Crater

The human history of Meteor Crater is itself interesting. The impact probably occurred before the coming of man to the Western Hemisphere, but the meteorites themselves apparently held some special significance to the early dwellers in the region for two separate meteorites have been found buried in small graves, one 40 kilometers (25 miles) northwest of the crater and the other 88 kilometers (55 miles) to the southwest. The latter was wrapped in a feather cloth, as was customary for the burial of children.

By the 1870s the crater was variously known as Franklin's Hole, Coon Butte or Crater Mountain. In March 1891 a prospector collected some of the metallic material for assay hoping to have made a rich strike. The sample found its way to A. E. Foote, a mineral dealer in Philadelphia who recognized the sample for what it was, a meteorite. He visited the locality in June of that year, collected more material and noted the crater, but did not speculate on its origin. As is customary, he named the meteorite Cañon Diablo (later changed to Canyon Diablo) for the nearest Post Office, in this case a train stop on the Sante Fe Railroad 16 kilometers (10 miles) northeast of the crater. Canyon Diablo is a deep, narrow canyon incised into the Colorado Plateau about 5 kilometers (3 miles) west of the crater (Figure 14–5).

G. K. Gilbert, a highly respected geologist in the U.S. Geological Survey, visited the crater in the summer of 1892. Despite the presence of considerable meteoritic material he rejected the impact hypothesis in favor of a volcanic steam explosion (phreatic eruption) for the formation of

Figure 14-5. Canyon Diablo today, with Meteor Crater in distance. The road crossing is **Interstate 40**. The canyon is a good example of incised meanders.

the crater. This is all the more ironic for it was Gilbert who seriously advocated a meteorite impact origin for the craters on the moon.

The disfavor paid to the impact hypothesis did not, however, find universal acceptance. In 1903 D. M. Barringer and co-workers filed a claim on the crater and established the Standard Iron Company, whose goal it was to locate the main mass of the meteorite which they believed to be buried somewhere beneath the floor of the crater. By 1909 28 holes had been drilled to a maximum depth of 250 meters (825 feet). No meteorite was found but the crushed nature of the Coconino Sandstone was well documented.

In 1909 Barringer read a paper before the National Academy of Science, strongly advocating the meteoritic origin of the crater, but the scientific community either was not swayed from Gilbert's hypothesis or else remained silent. Hypothesizing that the meteorite had hit with a low-angle trajectory, and might be beneath the crater rim rather than beneath the floor, a final drilling was begun on the south rim in 1920. In 1923 the drill became lodged in what was thought to be the main mass of the meteorite at a depth of 412 meters (1360 feet). Next a shaft was sunk to reach the meteorite, but this had to be abandoned at 225 meters (743 feet) depth due to uncontrollable groundwater which flooded the shaft. Financing had become increasingly difficult and the project was abandoned.

D. M. Barringer died in 1929 but the Barringer family maintained the lease on the property. Today they operate the crater as a public attraction through Meteor Crater Enterprises, which maintains the fine museum at the crater rim and supports scientific investigations in meteoritics at the crater and elsewhere in the world.

Scientists continued to speculate on the origin of the crater, and concerned themselves with calculations of the size of the supposed meteorite. Most of these papers were theoretical in nature, and few of the investigators had even been to the crater for field observations. Nevertheless the meteoritic hypothesis gained favor.

Then in 1946 an extraordinary man and his family appeared on the scene, brought with them a lifetime collection of meteorites from all over the world, leased a building on Route 66, 9 kilometers (5½ miles) north of the crater, opened their own museum, and began first-hand investigations around the crater.

The man was Harvey H. Nininger. As a biology teacher at McPherson College in Kansas he became intrigued by meteorites in 1923. The witnessing of a fireball (a particularly bright meteor shower) in November of that year led to a search which involved painstaking interviews of many persons who had also witnessed the event. Most reports were in error as to the direction of flight and the apparent landing spot but a probable impact area was ascertained. No meteorites from this display were ever found. However, due to Nininger's public lectures, discussions and newspaper articles, two other meteorite finds were made, both discovered by farmers while plowing their fields. Both were purchased by Nininger.

As his interest grew, so did the realization of the general ignorance about these objects from space. Several large collections existed, such as those at the Smithsonian Institution in Washington and the Field Museum of Natural History in Chicago, but only the mineralogy of these specimens had been described, and there seemed not much else that needed to be known about meteorites. Nininger was drawn into the vacuum. He decided to make a career of the collection, trading and the study of meteorites. Having a wife, Addie, and two children meant that an income was needed, so he held onto his teaching job, and devoted as much time as possible to enlisting public support in his search through lectures and to visiting known meteorite localities.

The plains states proved to be fertile ground, for in most places no hard bedrock is exposed, and any stone found in a field is an oddity. Meteorites were found holding down lids of cisterns and as doorstops, and so the collection grew. Following a highly successful expedition to Mexico in 1929 and the purchase and resale of a large meteorite in 1930 which netted a profit of $2000, Nininger resigned his post at McPherson, loaded the family into the 1929 Chevrolet, and headed for Denver to devote full time to meteorites.

The depression years had their highs and lows. Always on the verge of bankruptcy, the Niningers were sustained at times by patrons and friends, at times by selling portions of the ever-growing collection. Seldom if ever did support come from government or private granting foundations, which chose to reward only "legitimate" scientists in their ivory towers of academia. Then after World War II, Harvey and Addie moved their collection into an abandoned stone house north of Meteor Crater, opened their doors to the motoring public on Route 66, and Harvey began investigations on the crater. In 1949 the highway was rerouted and the attendance at the museum dropped off drastically. Again faced with financial collapse, they moved to Sedona in 1953 where they opened the American Meteorite Museum. At that time more than half of the meteorites discovered in the United States in the preceding 50 years had been found by Harvey Nininger.

The stream of visitors to the Sedona museum never flowed strongly during the summer and amounting to only a trickle in the winter, and the Niningers began considering the sale of their lifelong collection. They had been encouraged over the years by colleagues to sell the collection intact, rather than dividing it into smaller portions. The valuable collection would be more difficult to sell than would individual meteorites, but the British Museum of Natural History indicated an interest in buying a "vertical split," pieces cut from most of the meteorites. Before the funds had been raised, both the Smithsonian Institution in Washington and Arizona State University initiated efforts to obtain the collection. But the British tenured a cash offer first and on June 13, 1958 bought 21 percent of the Nininger Collection for $140,000, though it was not without some reluctance that Harvey and Addie let the collection leave the country. After carefully sawing and packing the specimens, the Niningers left on a grand tour of the Far East, of course with many meteoritic stops. Then in 1960 funds from the National Science Foundation, an anonymous benefactor and the A.S.U. Foundation enabled Arizona State University to purchase the remainder of the collection for $275,000. The Nininger Collection, which is the foundation of the largest University-owned meteorite collection in the world, is curated by the Center of Meteorite Studies at Arizona State, where several museum displays can be seen in the chemistry and geology buildings. The Center conducts ongoing research on the collection and makes specimens available to qualified scientists around the world. In addition the Niningers have established a prize, awarded through the Center of Meteorite Studies for the best original paper each year by a college undergraduate on the subject of meteorites.

The Canyon Diablo meteorite from Meteor Crater is not the only one found in Arizona. More than 30 others have been found including a ring-shaped iron meteorite weighing 688 kilograms (1520 pounds) which was used as an anvil by the Spaniards in the Tucson pueblo (Figure 14–6). The only observed fall in the state was that of the Holbrook meteorite on July 19, 1912.

The combined weight of all of the Canyon Diablo meteorites in public collections is about 11.5 tons. Estimates on the size of the impacting body which created Meteor Crater range from 5,000 to 5,000,000 tons depending upon the estimate of the meteorite's velocity, which of course is not known. The single, large mass that was so ardently sought by the Barringers will not be found for it is now thought that the meteorite exploded upon impact from the violent shock wave which it created. Much of the meteorite was probably volatilized and much remains in small pieces disseminated throughout the shocked debris beneath the crater. Meteor Crater has been likened to a crater produced by a 1.7 megaton nuclear explosion detonated at a depth of 20 meters (67 feet). It was used as a training ground for the lunar astronauts during the 1960's and continues to delight and intrigue scientists and public visitors alike.

Figure 14–6. Tucson meteorite, an iron, ring-shaped meteorite used as an anvil by the Spaniards. Photo courtesy of Center for Meteorite Studies, Arizona State University.

CHAPTER **15**

ECONOMIC GEOLOGY

Mining Arizona's Minerals

It can be said that mining is at the very foundation of our civilization, for the raw materials for most of our manufactured products have their origin in the ground. All else originates from the biosphere, for example, lumber, cotton and wool. These resources are renewable if managed properly; however the minerals and energy we take from the ground are nonrenewable. The processes responsible for their origin have been operating sporadically throughout the history of the earth and are operating today, but their slowness is measured in units many times greater than the span of the human race.

Mining has figured greatly in the economic prosperity of our nation, particularly since the west was opened up. The history and development of Arizona is largely a story of prospectors and mineral strikes, of miners and entrepreneurs, and finally great consolidated mining companies, which laid the economic foundation of the state.

The first mining in the state was done by the native American population, as much as 1000 years prior to the first contact by the Europeans. Turquoise artifacts have been found in the famous Snaketown site of the Hohokam, located southwest of Chandler on the Gila River Indian Reservation. These artifacts have been dated at between 500 and 700 A.D.

Prehistoric turquoise quarries have been found in the Dragoon Mountains, at the juncture of Canyon Creek and the Salt River, and at Mineral Park near Kingman. At the latter site are numerous pits and tunnels which are up to 7.5 meters (25 feet) deep. Stone hammers and mauls were used to break the rock free. There also is evidence that the Indians used fire in their quarrying operation. The rocks were heated and then quenched by throwing water onto them, causing pieces to break loose by the sudden contraction. Since turquoise turns white when heated much good material must have been lost in this process. Turquoise was highly valued by the Indians of the Southwest and Mexico, so active trading of this gem stone occurred throughout a wide area.

Exploration for metallic minerals in Arizona began soon after the Spanish conquest. In about 1530 an Indian appeared in the court of de Guzman, governor of New Spain, with tales of the Seven Cities of Cibola somewhere to the north, where such richness existed that the streets were paved with silver and gold. Such a tale might have been dismissed outright were it not for the arrival of Cabeza de Vaca, an explorer who had been shipwrecked on the coast of Florida and walked all the way to Mexico City, bringing with him a similar story of fabulous cities of gold, which he had heard about on his trek.

Immediately an expedition was organized under the direction of Estevan. The group made its way to Zuni country, mounted an assault on a pueblo, but were massacred by the inhabitants. Friar Marcos, the padre with the party, and the few others who did not participate in the battle, returned to Mexico with reports that Cibola had been found. A second, stronger expedition, this time led by Coronado, left Nayarit, Mexico, in 1540 to secure the riches for Spain. The party traveled through the White Mountains to the pueblo at the present location of Zuni, New Mexico. After a hard battle they defeated the Indians and triumphantly entered the city to find only adobe buildings and dirt streets. Not to return home empty-handed, the party followed a wild goose chase which led as far as Kansas and took two and a half years, but finally they gave up and returned to Mexico.

The first mineral discovery in Arizona was made by the explorer Espejo who located a silver deposit in the Verde Valley during an expedition through New Mexico and Arizona in 1582–83. He was a good enough economic geologist to know, however, that the deposit was not rich enough to make it profitable to pack the ore back home to southern Mexico, and so it was not exploited at the time. This was probably the deposit that was eventually worked as the United Verde Mine at Jerome.

An important figure in the early history of Arizona was Father Kino, a Jesuit who established a line of missions in Sonora and southern Arizona beginning in 1691. Within the present state boundaries these included San Xavier del Bac, Tumacacori, Guevavi, and Santa Cruz.

Father Kino was interested primarily in bettering the lives of the Indians in the area and promoted improved agricultural practices and trading. Though he was not directly involved in mineral development, several mines were opened within the area that he had settled.

In 1736 a Yaqui Indian discovered the Planchas de Plata deposit near the town of Arizonac, southwest of Nogales. This was a very rich native silver deposit which stimulated much prospecting in the general area. But more importantly perhaps is that the town eventually lent its name to the Arizona territory and state.

Following the collapse of the Spanish Empire, little activity occurred in the area until after the Gadsden Purchase in 1854, which secured the lands between the Gila River and the present day Mexican border for the United States. Prospectors exploited mainly placer deposits of gold and silver, with high grade pockets of ore. The copper mine at Ajo which had first been worked by the Spaniards was reopened in 1854, but in order to get the ore to a smelter it was packed by mules to the Colorado River where it was loaded onto boats and shipped to Swansea, Wales, where it was sold for $360 per ton and processed into the metal.

A constant threat at this time was a raid by Indians, particularly the Apaches, so that when federal troops were withdrawn at the outbreak of the Civil War, travel became impossible except in large, armed groups. One particular success came in 1863 when Henry Wickenburg discovered a thick vein of gold west of the present day town that bears his name. The prospect became the Vulture Mine (Figure 15–1), the richest of its day, and much more gold was recovered from the gravels of the Hassayampa River and its tributaries such as Weaver and Antelope Creeks near Congress.

Following the Civil War, opportunists flocked to the Arizona Territory. Roads were improved and reliable trade became established. The entire state was scoured by prospectors and most of today's great copper deposits were discovered, though at this time they were of little economic importance.

In the early 1880's, railroads were built across the territory, connecting with lines both to California and the East. Supplies became cheaper as did the cost of getting ore and metals to their markets. The silver market, which had been booming, collapsed in 1893 when silver ceased to be used for the monetary standard.

Figure 15-1. Headframe and mill at Vulture Mine, Wickenburg.

However, copper had been gaining in importance. The electric age was dawning and with it came the demand for a cheap, tough conductor to carry electric power. In 1886 the first copper concentrator in Arizona was built at Morenci, soon to be followed by one at Clifton, so that increasingly lower grade ores could be exploited. In 1888 for the first time the revenue generated by copper exceeded that of gold and silver, $5,300,000 versus $3,000,000. And the difference continued to grow. In 1907 copper production grossed $50,000,000 and in 1910 Arizona passed Montana as the nation's leading producer of copper, a position which it has retained ever since.

The economics of mining are basic. One must be able to sell the metal he has produced for more than it cost to produce it, or else, in our capitalist system, he goes broke. Some countries which compete in the world copper market subsidize their production in order to stimulate exports and secure foreign exchange to buy import products. The effect in recent years has been to undercut domestic U.S. prices and at times cause economic trauma in the copper industry. The costs of a mining operation include removal of the ore from the ground. *Ore* is any rock that is saleable at a profit. Costs also include removal of waste rock, or *gangue*. If the mine is underground there are the costs of stabilizing the walls and roofs, of ventilation, and sometimes pumping out water or pumping in air.

Once the ore is removed from the ground there are the costs of processing it. These include crushing the rock, milling it, smelting it, and refining it. The finished product is then sold to manufacturers who will transform it into a myriad of products upon which we depend.

In Arizona, copper is king, and the biggest of the copper mines are of the *open pit* variety. These are simply large holes dug into the ground with sides at an angle permitting a road to spiral up from the bottom, so that loaded trucks may haul the ore to the processing plant (Figure 15-2).

Figure 15–2. Open pit copper mine and plant at Baghdad.

Open pit mines are the cheapest to operate because the only costs are those of removing the ore: blasting, steam shovels, trucks, and people who operate them. The goal of an open pit operation is to move as much rock as possible, permitting exploitation of lowgrade ores. Trucks often weigh 85 tons and carry an equal amount of ore. In Arizona the lowest grade material that can be classified as ore must contain at least 0.5% copper in the rock.

Underground mines are more expensive to operate, so they must recover higher-grade ores. They also are more efficient in that they can follow high-grade seams or veins without digging large volumes of gangue that would have to be removed in an open pit operation. Expenses of an underground mine include the price of timbers and bolts to shore up the roof and prevent cave-ins, costs for ventilation and sometimes pumping, electricity for lighting, trains and tracks.

The layout of a mine is dictated by the geometry of the ore body. It is centered around a vertical tunnel called the *shaft* (Figure 15–3). Usually both the miners and the ore ride up the shaft to the surface on an elevator called the *skip;* however, the two do not often share the ride. Side tunnels called *drifts* are sunk into the ore body from the shaft and outfitted with narrow gauge railroad tracks, electric engines and ore cars for transporting the ore from the working areas to the shaft.

One common mining technique is called *stoping.* The area to be mined, usually at the end of a drift, has blasting charges set in the ceiling, which when detonated cause a portion of the roof to fall in. The loosened rock is then loaded onto the cars and removed. Another charge is set, more rock falls in from the roof and the process is repeated. This ever enlarging room is known as the *stope* (Figure 15–3). Recall that the term "stoping" is also applied to a style of igneous

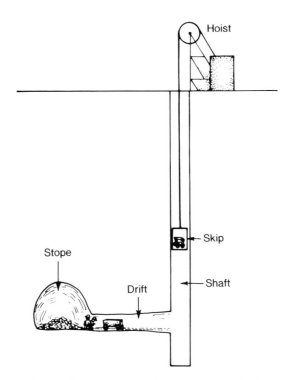

Figure 15-3. Diagram of an underground mine.

intrusion (Figure 6–1). The limits to the size of the stope are dictated by the strength of the rock and its ability to span an arch over the room. Highly fractured rock of course is more susceptible to collapse. In order to remove the maximum amount of ore and still maintain the strength of the mine, stopes are sometimes backfilled with *tailings,* the powdered waste rock stripped of its copper in the processing plant.

After the ore is removed from the mine it goes to the processing plant. The type of process used to remove the copper from the rock depends on the type of ore being mined. The two types of copper ores are 1) those containing primary sulfides, such as chalcopyrite ($CuFeS_2$), chalcosite (Cu_2S) and bornite (Cu_5FeS_4), and 2) those containing secondary minerals, such as chrysocolla ($Cu_2H_2Si_2-(OH)_4$), malachite ($Cu_2CO_3(OH)_2$), and cuprite (Cu_2O). In both cases the first step is to crush the ore into a fine rock powder. The first crushing is usually done in a *gyratory crusher,* a tremendous conical-shaped device with a steel spindle in the center that has a skewed rotation (Figure 15–4).

With sulfide ores the rocks are often further ground in a *ball mill* or a *rod mill.* The ball mill is filled with ore and steel balls which are shaken violently to pulverize the rock, whereas the rod mill is a large rotating cylinder that contains large steel rods that roll over one another and grind the rock. Water is added to the rock powder at this step of the operation producing a sludge. By now the individual grains of sulfide have been broken loose from the silicate minerals and are ready for the flotation process.

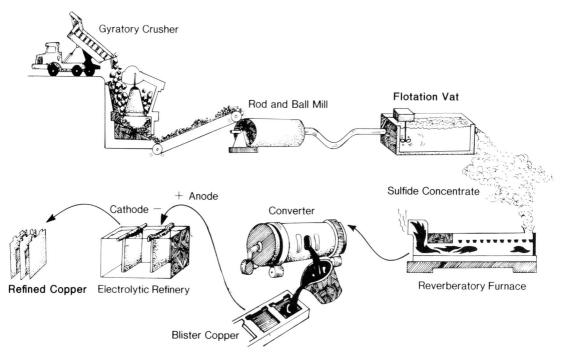

Figure 15–4. Stages in processing of sulfide copper ore.

When certain oils are beaten in water they produce a froth to which the sulfides, but not the silicates or oxides, will adhere. The sludge from the mill is diluted with water, oil is added, and the whole mixture is whipped. The sulfide-containing froth rises to the top of the flotation vats where it is skimmed off. This concentrate now contains 15% to 30% copper. The unwanted rock powder settles to the bottom of the vat and is removed from time to time. This waste material, called *tailings*, is then usually dumped in orderly piles, which are a familiar sight around large mining operations.

The copper concentrate is next taken to the *smelter*. Fluxing agents are added and the mixture is fired in a *reverberatory furnace* until it melts. Slag and a mixture of copper, sulfur and iron, known as *matte*, are produced. These two products separate in the melt because of different specific gravities, and the slag is drawn off. The matte is introduced to a second furnace called a *converter* where the iron and sulfur are removed. The sulfur combines with oxygen to form sulfur dioxide which escapes as a gas. In times past many companies simply released the sulfur gas, which when mixed with water in the atmosphere would produce sulfuric acid, so-called "acid rain," which causes much damage to the vegetation surrounding the mine. Recently most companies have installed sulfuric acid plants that remove SO_2 from the smelter exhaust and utilize it in the treatment of copper oxide ores to increase the recovery of copper from the mine.

Blister copper, which is 99.5% pure, is produced in the converter furnace. It is poured in the molten state into molds and then taken to an *electrolytic refinery* for further purification. The molded sheets of blister copper are hung in a solution of dilute sulfuric acid and copper sulfate, adjacent to thin sheets of very pure copper. Electric current is fed into the blister copper which acts as an *anode,* or the positive electrode, of a battery. As the current passes to the pure copper *cathodes,* or negative electrodes, the copper is dissolved at the anode and reprecipitated at the cathode. During this transfer most of the remaining impurities fall to the bottom of the vat. These sometimes include gold and silver, which can be a very lucrative byproduct for a mine. The copper deposited at the cathode is 99.99% pure and is ready for industrial applications requiring the purest of copper.

The process for extracting copper from oxide ores is simpler (Figure 15–5). Once the ore is crushed it is put in large *leaching vats* lined with lead, sulfuric acid is poured over it and left for about a week. The sulfuric acid dissolves the copper oxides, turning a bright blue in the process, and then the copper-charged acid is piped into an electrolytic tank similar to that used for sulfide ores. This time, however, the anode is made of lead and the cathode, copper. When electric current is passed through the system the copper in solution is plated out onto the copper cathodes which are then ready for the industrial market.

Not quite all of the copper is plated out of solution by electrolysis so as a final recovery step the acid is poured over scrap iron where the last of the copper is precipitated on the iron. The vats containing the scrap iron are called *iron launders.* This copper along with the remaining iron is then taken to a smelter where the copper is separated from the iron in the manner already described.

Arizona is blessed with an exceptionally large concentration of copper deposits. These were emplaced mainly between the late Cretaceous and middle Tertiary, always in association with porphyritic intrusive rocks of intermediate composition, mainly granodiorite. The common term "porphyry copper" is derived from this association. Porphyry copper deposits characteristically are enormous, with individual ore bodies measured in the millions or sometimes billions of tons. On the other hand they are very low grade. At present the economic limit for mineable ore in Arizona is about 0.5% copper, ten pounds per ton of rock, but vast areas with lower percentages also exist, awaiting a rise in prices or a breakthrough in technology to make them mineable.

The primary copper minerals are generally sulfides like chalcopyrite and chalcocite. These are precipitated in the rock from hot water solutions (*hydrothermal solutions*) which contain dissolved copper and sulfur as well as a number of other elements, including gold and silver. The

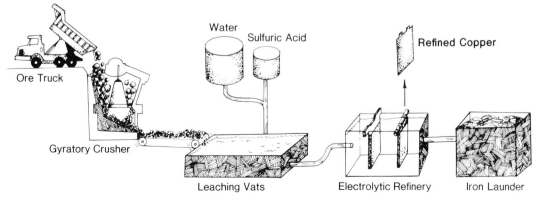

Figure 15–5. Stages in processing of oxide copper ore.

various elements contained in the hydrothermal solutions are concentrated in the last uncrystallized residue of the magmas which solidified to form the porphyritic granodiorite. This happens because, as a magma crystallizes, the minerals that grow in the melt are mainly silicates containing the eight most abundant elements (see Chapter 1). The metallic elements like copper and gold, and certain volatile elements like sulfur, chlorine, and fluorine do not combine with the silicates and so are successively enriched in the diminishing magma as it undergoes crystallization.

The origin of the water in the hydrothermal solutions is not certain. On the one hand water (H_2O) is concentrated in the final portion of the magma, and may be enough to produce a hydrothermal solution. On the other hand, ground water sometimes circulates to a great depth, and may mix with the residual magma to produce the ore-bearing fluids which ascend and deposit their minerals in the surrounding rock.

Often times the intrusive rocks containing the ore minerals have been highly fractured or brecciated, with the cracks and open spaces serving as passageways for the rising hydrothermal solutions, and as channels for concentrated precipitation of the ore. However, the ore minerals are also dispersed throughout the rock by replacement of some minerals by others.

When the hydrothermal solutions move out of the igneous intrusions into the intruded country rock, they may cause metamorphic reactions and ore deposits known as *skarns,* which sometimes are quite high grade. These are particularly well developed where the fluids react with limestone.

The porphyry copper and skarn deposits are of a *primary* type. If they are brought to the surface and exposed by erosion, reactions of a *secondary* type will occur resulting in the oxidation of the sulfide minerals. Some of the copper will remain in the rock near the surface as oxides, while the rest will be washed downward by ground water percolation and concentrated beneath the zone of oxidation, usually beneath the water table. This *secondary enrichment zone* may be enriched just enough to make the difference between an economic and an uneconomic deposit, or it may be extremely rich. The copper left in the *oxide zone* may also be economically exploitable by the methods discussed earlier.

Energy Resources of Arizona

Everyone is painfully aware of the increasing demands of society for energy supplies. The conventional source of energy to which we have become accustomed is petroleum and natural gas, but due to dwindling supplies worldwide, other sources such as coal, oil shale and tar sand, nuclear fuel and geothermal energy are becoming increasingly attractive. The greatest energy resource in Arizona is coal, but significant amounts of uranium, oil and gas have been produced. The development of geothermal resources is just beginning. Each of these energy resources in Arizona will be discussed briefly in the following pages.

Oil and Natural Gas

Introduction—Arizona's known oil and gas supplies are meager, and known fields are rapidly becoming depleted (Figure 15–6). However exploration for new sources has accelerated during the late 1970s and at this writing a massive leasing program has just been completed based on the theory that nearly all the Basin and Range Province may be underlain by potentially productive Mesozoic and Paleozoic sediments beneath an "overthrust sheet" of Precambrian rocks. An example of an imaginative exploration effort, the concept is based upon an interpretation that the

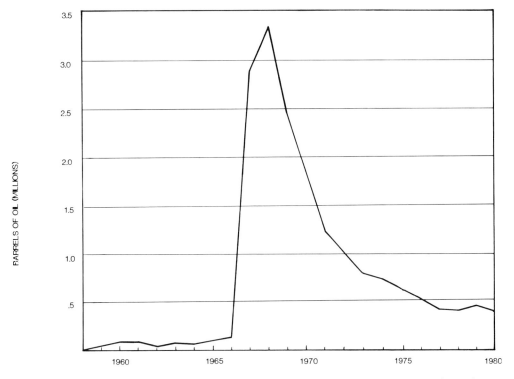

Figure 15–6. Production graph of oil and gas in Arizona. Modified from Hale, M. E. (Editor) 1976. Arizona, Energy, A framework for decision, University of Arizona Press.

extremely productive thrust belt in the Rocky Mountains of Wyoming and Utah may extend through Arizona to connect with similar structures in Mexico (Figure 15–7). Recent studies in western Aaizona have demonstrated thrust faulting in the Rawhide, Buckskin, Harquahala and other mountain ranges of Yuma and Mohave Counties. A well was drilled (1980) to 5455 meters (18,000 feet) depth, 23 kilometers (14 miles) south of Florence to test the theory (Figure 15–8). That test proved to be unsuccessful but additional deep tests are planned to explore the overthrust and other theories for oil and gas accumulations in Arizona.

Before we discuss the known oil and gas occurrences and production in Arizona, we should first consider the geological conditions necessary for their accumulation. The prerequisites are: 1) source rock, 2) reservoir rock, 3) trap rock and structure.

Source rocks are those which contain a high proportion of organic matter. The organic matter was incorporated in the sediments at the time of deposition and probably consisted primarily of algae, and single-celled and multi-celled animals that were buried in the mud of the sea floor. By far, the greatest proportion of known oil occurrences in the earth have been formed in marine environments, however extensive deposits of oil shales are known to have formed in lacustrine sediments, probably due primarily to the algae that flourished in such environments. The mere occurrence of large amounts of organic matter is not adequate to form oil and gas. The material must be buried deeply enough for maturation to occur. The conditions necessary for maturation

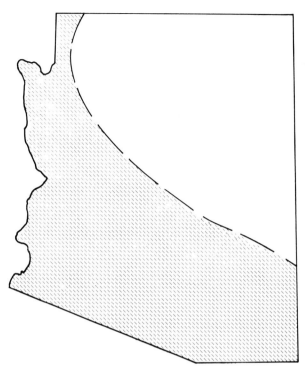

Figure 15-7. Map showing trend of the overthrust belt as it has been projected across Arizona.

Figure 15-8. Drilling rig on location of 18,000 foot oil test, south of Florence.

seems to be heat (temperatures above 60°C) and time (several millions of years). When both conditions are met the volatile and liquid fractions are distilled off and slowly migrate through minute pores and fractures in the rock in the direction of lower pressure—generally upward toward the earth's surface. As the oil and gas migrates through the layers of superposed strata, its movement is facilitated into and through those layers with large and interconnected pores and/or fractures. Such porous rocks, called *reservoir rocks,* are normally well-sorted sandstones or porous limestones, but may be other kinds of rocks such as fractured or weathered shales, cherts, volcanic or other igneous rocks, with secondary porosity development.

In order for the oil and gas to be concentrated in producible quantities in the reservoir, a *trap rock* and *structure* is necessary to prevent its migration and dispersal. The usual traps are formed by shales, evaporites or other impermeable rock layers that are folded in anticlines (Figure 15–9), displaced against reservoir rocks by faults (Figure 15–10), overlie reservoir rocks in unconformities (Figure 15–11), or enclose reservoir rocks in stratigraphic pinch-outs (Figure 15–12).

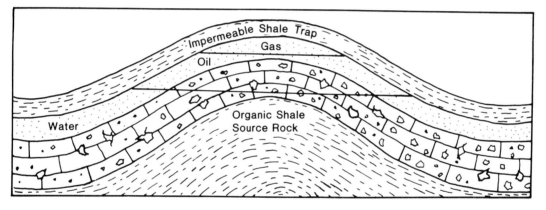

Figure 15–9. Diagram of anticlinal oil and gas trap.

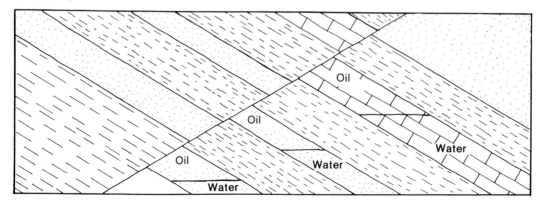

Figure 15–10. Diagram of faulted oil and gas trap.

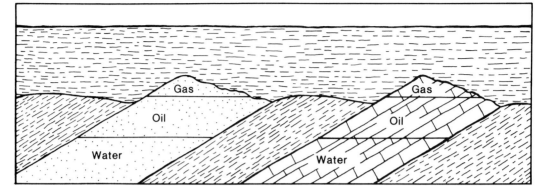

Figure 15-11. Diagram of oil and gas trap at an unconformity.

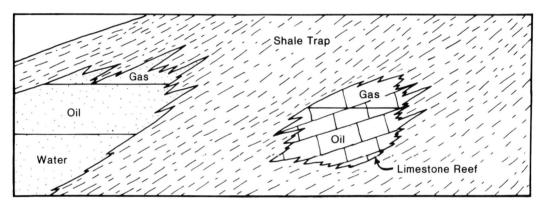

Figure 15-12. Diagram of a stratigraphic pinch-out oil and gas trap.

With these general concepts of oil and gas occurrence in mind, let us discuss the oil and gas production and potential in Arizona.

Occurrence—Oil and gas can only be generated where thick sequences of sediments have accumulated in depositional basins. The concentration of these fluids can only occur where favorable structural and/or stratigraphic traps have been formed. In Arizona, two regions have been identified where these favorable conditions are known to exist—the Plateau Province of northern Arizona and the *Pedregosa Basin* in southeastern Arizona.

All production of oil and gas in Arizona has been from the Plateau Province of northern Arizona (Figure 15–13). The area was located on a continental shelf throughout the Paleozoic which separated the Cordilleran Geosyncline in Nevada from the emergent Transcontinental Arch in New Mexico. The Paleozoic sedimentary rocks record a fluctuating shoreline resulting in a discontinuous sequence of marine sediments interrupted by unconformities and/or nonmarine sediments. All the Paleozoic systems except the uppermost Permian, are terminated toward the New Mexico border either by onlap or erosional truncation along the western margin of the

Defiance Uplift and thin southward over a positive area in central Arizona. Most of the Paleozoic systems thicken toward the Paradox Basin in southeastern Utah in the Four Corners region.

During the Triassic and Jurassic the area was buried by predominantly nonmarine sediments, but was submerged beneath the shallow Western Interior Seaway in Late Cretaceous time (Figure 8–3), only to become emergent again in latest Cretaceous time.

The Colorado Plateau in northern Arizona is a relatively undisturbed structural province but has been downwarped into two major structural basins—the Black Mesa Basin and the Paradox Basin. The Paradox Basin barely extends into Arizona but all of the Arizona oil and gas production is confined to it. The Black Mesa Basin is bounded on the east by the Defiance Uplift (positive since Precambrian) on the north by the Monument Upwarp, on the west by the Kaibab Plateau and on the south by the Mogollon Slope, all Laramide in age (Figure 8–3). Several folds and faults with predominantly N–S or NW–SE trends cross the region, each of which create potential traps for the oil and gas that may have been generated within the sedimentary deposits.

The greatest exploration activity for oil and gas in southern Arizona has been concentrated in Cochise County in the extreme southeastern corner of the state. The Pedregosa Basin (Figure 10–11) is characterized by a thick (more than 600 meters; 2,000 feet) sequence of Pennsylvanian rocks with distinct deep marine basin facies of limestone and mudstone, flanked by marginal

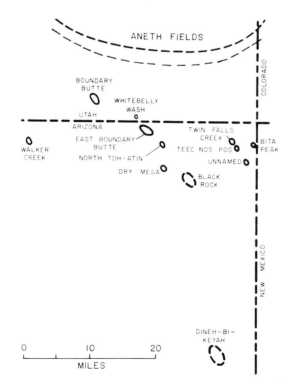

Figure 15–13. Map of oil and gas fields in Arizona. Modified from Conley, J. N., 1974. Review of the development of oil and gas resources of Northern Arizona, in Geology of Northern Arizona, *Geological Society of America Guidebook.*

porous dolomite (reef?) and shelf carbonates. The similarities between this basin and the highly productive Permian Basin of New Mexico and Texas have attracted attention to its potential. The additional thickness of Permian and Lower Cretaceous rocks known to be in the basin increase its potential as a petroleum province. Over 40 petroleum exploration wells have been drilled in the Arizona portion of the Pedregosa Basin resulting in some encouraging shows but no commercial production.

Structural complications have been imposed on the Pedregosa Basin by Laramide thrusting and wrench faulting, but the most severe modifications have resulted from Mesozoic and Cenozoic igneous intrusions and Late-Tertiary Basin and Range extensional faulting.

Production—All oil and gas production and known reserves in the state are confined to Apache County on the Colorado Plateau (Figure 15–13). Forty-five wells have produced about 17 million barrels of oil in this area of 10 fields, the largest of which is the Dineh-bi-Keyah Field with 25 wells completed in a Tertiary sill intruded into Pennsylvanian marine strata. Fifteen of the additional 20 wells produced from Pennsylvanian strata, 4 from Mississippian, and 1 from Devonian. Structurally, the Dineh-bi-Keyah Field is on the flank of the Defiance Uplift and all others are on the same structural trend in the southern margin of the Paradox Basin.

Potential—In the Plateau Province southwestward from the Four Corners area, the Pennsylvanian-Permian strata grade laterally to redbed and limestone facies in the Black Mesa Basin, where the Devonian and Mississippian marine strata appear to be most favorable for oil and gas. These strata wedge out to the east against the Defiance Uplift and to the south against the Mogollon Slope, creating favorable conditions for stratigraphic traps. In rocks of Permian age the Fort Apache member of the Supai Formation, the DeChelly sandstone and the Coconino Sandstone have some potential for shallow production. To the west in Coconino and Mohave Counties an eastward regional thinning of Cambrian-Pennsylvanian strata and local thinning over the Kaibab positive, creates potential stratigraphic and structural traps.

Several other large, and largely untested, areas of the Plateau are known to contain Paleozoic rocks with some potential, such as the Coconino Plateau south of the Grand Canyon, with Devonian and Mississippian potential; northwestern Arizona north of Grand Canyon, with a thick lower Paleozoic section and Paleozoic rocks buried beneath the White Mountain Volcanic Field.

In southeastern Arizona, most exploratory drilling has been done in the Pedregosa Basin where encouraging shows of oil and gas have been encountered. The Horquilla Limestone (Pennsylvanian) and Martin Formation (Devonian) have the highest potential in southeastern Arizona based upon the best shows in the Horquilla and good reservoir characteristics in the Martin Formation. Lower Cretaceous sedimentary rocks (Bisbee Group) up to 3600 meters (12,000 feet) thick were deposited as a delta complex on the margin of a sea that extended to the southeast into Mexico. It contains a limestone unit, the Mural Limestone, ranging from 90 to 240 meters (300 to 800 feet) thick, which has favorable reservoir characteristics and some shows of oil and gas.

Many of the wells in southeastern Arizona have been drilled on Basin and Range horsts where reservoirs tend to be flushed by ground water, and the best prospects lie in the graben where chances for oil and gas preservation are better. None of the wells have tested the deeper parts of the basins in Cochise County where thick sections of Paleozoic rocks should be encountered at depths of below 3030 meters (10,000 feet). Future exploration should also evaluate the potential of the Lower Cretaceous objectives in favorable locations.

There has been very little drilling for oil and gas in the Basin and Range Province of southwestern and west-central Arizona. However, as mentioned previously, it is an area of active geophysical exploration and leasing activity based on the theory that the "overthrust belt" crosses Arizona in an arcuate pattern from the northwestern to the southeastern corners of the state (Figure 15–7). The theory is that the predominantly Precambrian igneous and metamorphic rocks exposed at the surface in that area are part of a thrust plate that has been pushed several tens or hundreds of kilometers from the west, over a thick sedimentary basin of Paleozoic and Mesozoic rocks which would likely contain oil and gas. The first attempt to drill through the thrust plate into sedimentary rock below was unsuccessful as the hole bottomed in granite at 5,455 meters (18,000 feet) below the surface. Several other wells have been drilled to test the theory, but no discoveries have yet been made that can be related to overthrusting.

In addition to the overthrust belt potential in southern and western Arizona, there is potential for oil and gas occurrences in Cenozoic or Paleozoic strata preserved in Basin and Range grabens. Recent discoveries of oil and gas accumulations in the Great Salt Lake area of Utah and Railroad Valley in Nevada demonstrate that potential. As an example, the Eagle Springs Field in Nevada has produced over three million barrels of oil from Oligocene tuffs and lake beds which are buried beneath younger valley fill. The more recently discovered Trap Spring and Salt Lake Fields further demonstrate the potential for oil discoveries in the Basin and Range. Many similar grabens are known in southern and western Arizona which probably contain buried Paleozoic and Mesozoic sedimentary source rocks and Cenozoic volcanic and sedimentary rocks that have favorable reservoir characteristics.

Coal

Introduction—Coal is Arizona's most abundant fuel energy resource. It is restricted essentially to rocks of Cretaceous age, with the main reserves concentrated in the Black Mesa Field in northeastern Arizona (Figure 15–14). Several smaller deposits of only local economic significance occur at scattered locations in eastern Arizona, all Cretaceous except one in Late Paleozoic rocks exposed along the Mogollon Rim.

The most extensive coal reserves occur in Late Cretaceous rocks that have been preserved in a structural basin with considerable topographic relief (1818 to 2424 meters; 6000 to 8000 feet elevation) called Black Mesa. The coal beds crop out around the periphery and on the eroded top of the mesa, defining an areal extent of about 8192 square kilometers (3200 square miles). The coal interbedded with sandstones and shale of the Dakota Sandstone (Figure 11–11), Toreva Formation and Wepo Formation, which, along with the unproductive marine Mancos Shale and Terrestrial Yale Point Sandstone, form a combined thickness of 515 meters (1700 feet) (Figure 11–13). All these rocks dip toward the center of the Black Mesa Basin resulting in burial of the coals in the Dakota Sandstone (up to 515 meters; 1700 feet), the Toreva Formation (up to 303 meters; 1000 feet) and the Wepo Formation (between 98 and 242 meters; 325 and 800 feet). Total coal reserves beneath Black Mesa have been estimated at 21.25 billion short tons, with strippable coal within 39 meters (130 feet) of the surface at about one billion tons. The application of subsurface mining techniques below 39 meters (130 feet) would increase the recoverable coal from the area. Coal seams in the Dakota Sandstone average 0.6–1.2 meters (2–4 feet) thick with an observed maximum of 2.7 meters (9 feet) along the southwestern margin of the mesa. The thickest coal in the Toreva Formation is 1.8–2.1 meters (6–7 feet) thick in the northwestern rim of the mesa. The Wepo Formation contains the best quality coal in Black Mesa, and occurs nearer to

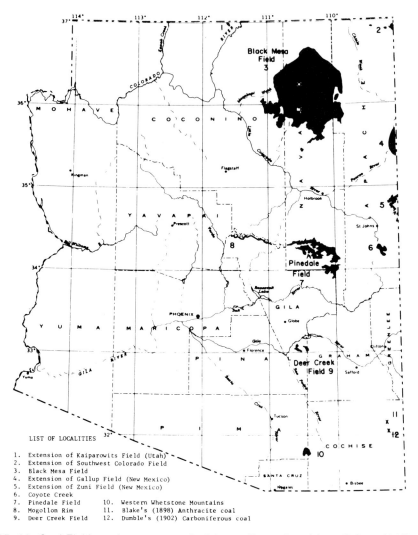

Figure 15-14. Coal Fields and occurrences in Arizona. Reproduced from Peirce, H. W. and J. C. Wilt, 1970. Coal; *IN* Coal, Oil , natural gas, helium and uranium in Arizona by Peirce, H. W., S. B. Keith and C. J. Wilt; Arizona Bureau of Mines Bulletin 182.

the surface than the Toreva or Dakota. It contains at least ten coal beds that individually exceed 1 meter (3 feet) in thickness. This formation is currently being mined by Peabody Coal Company at the northern margin of Black Mesa (Figure 15–15).

Production—Although coal mining on Black Mesa dates back to prehistoric times, large scale mining did not begin until 1970 when Peabody Coal Company started production on a 14,000 acre lease on tribal lands on the north side of Black Mesa. Since then they have been providing coal for two generating plants, the Mohave plant near Bullhead City, Nevada via a 275 mile slurry pipeline, and the Navajo plant near Page, Arizona via an 80 mile long railroad. About 40

Figure 15-15. Peabody coal mine in Wepo Formation, Black Mesa.

million tons had been produced at the end of 1980 and production is planned for about 12.5 million tons/year for approximately 32 years.

It has been estimated that 16 million tons of strippable coal underlie each square mile of the Peabody lease and that it contains the energy equivalent of 71 million barrels of crude oil per square mile.

Uranium

Uranium is the most abundant radioactive element in the earth's crust. It normally occurs only in trace amounts in plutonic and volcanic rocks, but under favorable circumstances may be concentrated enough to be extracted economically. It is becoming increasingly in demand as fuel for nuclear power plants in addition to military applications. Arizona is not a major producer of the metal now but has had significant production in the past and undoubtedly will again become a producer.

Occurrence—Numerous occurrences of uranium are known throughout Arizona with past production primarily from Triassic and Jurassic strata on the Colorado Plateau, but with current exploration and development increasing in the Basin and Range Province. This section will summarize the mode of occurrence, past production and current activity of uranium exploration and production in Arizona.

Uranium deposits in Arizona are of two general types; 1) *sedimentary deposits* and 2) *veins and fracture fillings*. The most abundant and generally most productive are the sedimentary deposits which are mainly in sandstone and conglomerate of continental origin. They consist of masses of rocks impregnated with uranium oxides, commonly in association with vanadium and sometimes iron, lead and zinc. The uranium content ranges from trace amounts to several percent, but the average grade of ore mined has been about 0.29 percent of uranium oxide.

The emplacement of uranium minerals in sedimentary rocks has resulted from post-depositional precipitation from groundwater solutions, in pore spaces or as replacement of grains, cement or fossil plant material. The most significant production has been from Mesozoic age terrestrial sediments laid down by slow moving, braided and meandering fresh water streams on deltas, alluvial plains or flood plains; or in lake basins.

The Colorado Plateau production in Arizona has come predominantly from two Mesozoic formations, the Triassic Chinle Formation (Figure 11–2) and the Jurassic Morrison Formation (Figure 11–10). The following brief discussion summarizes the occurrences and production of uranium in these Mesozoic systems.

Chinle Formation—Deposits in the Chinle Formation occur mostly in the basal 45 meters (0–150 feet thick) Shinarump Conglomerate (Figure 5–6) composed of sand and gravel that was spread by meandering streams over an extensive erosional surface cut into the underlying Moenkopi Formation. Many small and medium-sized deposits are in similar lenticular channel-filling sandstones in the lower part of the Petrified Forest Member. The uranium ore bodies are localized in conglomeratic sandstone that fills stream channels scoured into the Moenkopi Formation. It is probable that the uranium was precipitated from ground water flowing through the permeable channels in localities formerly occupied by plant material. Deposits in the Chinle Formation are concentrated in two areas of major production, one in Monument Valley and the other near Cameron in the valley of the Little Colorado River, with other smaller deposits in Chinle Formation outcrops elsewhere on the Colorado Plateau.

Morrison Formation—Uranium ore in the Morrison Formation in Arizona is essentially restricted to the Salt Wash Sandstone Member, which crops out in the vicinity of the Carrizo and Lukachukai Mountains in the extreme northeastern corner of the state. It consists of lenticular lenses of sandstone interbedded with mudstone. 85% of the ore mined from the Morrison Formation in Arizona has come from the Lukachukai Mountains.

Toreva Formation and Tertiary Basins—A few productive deposits have been found in fluviatile (stream-deposited) sandstone interbedded with carbonaceous siltstone in the lower member of the Toreva Formation on the northeastern margin of Black Mesa (Figure 11–12). However, the easily located surface exposures of uranium ore along exposures of productive Mesozoic rocks have probably been found, therefore more recent exploration has shifted to other objectives including Paleozoic rocks along the Mogollon Rim and Cenozoic rocks in the Basin and Range Province. The greatest industry activity has been concentrated recently in western Arizona (e.g., Anderson Mine, Yavapai Co.)(Figure 15–16) where the ore occurs in Tertiary lacustrine (lake-deposited) volcaniclastic sedimentary rocks in association with carbonaceous materials. The estimated reserves of Anderson Mine alone (50,000 tons U_3O_8) are greater than Arizona's cumulative production of uranium oxide. The ore emplacement is believed to be an early diagenetic event,

Figure 15–16. Anderson (uranium) Mine in southern Yavapai County.

resulting from the compaction and dewatering of uranium-rich volcanic lake sediments with precipitation of uranium caused by contact with a strongly reducing marshy environment. The ore occurs in several mineralized beds that are generally 1–3 meters (3–10 feet) thick, but locally range up to 11 meters (36 feet), and commonly reaches an aggregate thickness of 15 meters (50 feet). The percentage of uranium oxide in the ore averages about 0.06%.

Numerous occurrences of uranium in the Basin and Range Province of the southwestern half of Arizona have been reported. Most are associated with lake deposits consisting of interbedded sandstone, shale, mudstone, bentonitic clays, gypsum and volcanic ash or tuff. Carbonaceous material, silica and calcium carbonates are common associates. Considering the vast amount of Cenozoic sediments in the southwestern half of Arizona, it is possible that additional deposits will be found in the numerous basin areas of southern and western Arizona.

Other Modes of Uranium Occurrence

Diatremes and Breccia-pipes—These explosion and/or collapse-formed masses of fractured rock have been one of the major types of uranium occurrence in Arizona but due to small size and difficulty of exploration, they do not have the future potential comparable to the sedimentary deposits previously discussed.

The main occurrences of diatremes are in the Hopi Buttes and Monument Valley areas, and many of the uranium deposits are associated with infilling lacustrine limestone, sandstone and volcanic ash deposits. Most deposits are low grade and have produced only a few tens of tons.

The most productive deposits in breccia-pipe structures is the Orphan Mine on the south rim of the Grand Canyon, which has produced about 500,000 tons of good grade uranium ore. It and other breccia-pipes in the area originated by the collapse of solution caverns in the Mississippian Redwall Limestone, with the resultant collapse of the overlying strata as a breccia-filling. Several such structures are exposed, and many others must occur on the Plateau, offering the potential for additional discoveries.

Vein Type Deposits—Vein deposits that have been evaluated most thoroughly occur in the Younger Precambrian Dripping Spring Quartzite that crops out in the Central Mountain Region of Arizona. Production has amounted to only about 23,000 tons averaging 0.23 percent uranium oxide and has not been economically sound. However, it continues to be investigated as a host for low-grade deposits, and some production was recently achieved by solution mining. The mineralization probably occurred at the time of intrusion of Precambrian diabase sills in the Apache Group (Figure 6–2).

Numerous other small vein-type occurrences have been found in association with intrusive and extrusive rocks in the Basin and Range Province. The host rocks are mainly Precambrian granitic and rhyolitic rocks, but some have been found in Mesozoic intrusives and a few in Precambrian schist. The potential importance of such igneous and metamorphic occurrences of uranium is suggested by similar deposits in Alaska, the Basin and Range, the Rocky Mountains, in the Canadian Shield and in the Appalachians.

Geothermal

Geothermal energy is derived from the natural heat from the earth's interior. It is commonly brought to the surface by hot water moving up fault planes. Many hot or warm springs occur in the Basin and Range Province of southern and western Arizona, indicating the potential for geothermal resource development in those areas.

Investigation of Arizona geothermal resources began in 1971 when state and federal agencies, utility companies and private interests began geological and geophysical exploration for resources. During that year, a review of geothermal resources in Arizona listed 12 selected areas of thermal springs with temperatures ranging from 85°C to 40°C (185°F to 104°F). These thermal waters in the Basin and Range are closely associated with faults and have probably resulted from the cycling of surface waters. The occurrence of ground water in areas of relatively recent volcanism and faulting creates a potential for the development of geothermal energy. The U.S.G.S. circular, Assessment of Geothermal Resources of the United States, records eight identified hot water convection systems in Arizona, one with temperatures above 150°C (302°F) and seven with 90°C to 150°C (164°F to 302°F).

An investigation of geothermal energy resources in Arizona was begun in 1977 as a joint effort between the University of Arizona, Geosciences Department; the Bureau of Geology and Mineral Technology, Geological Survey Branch; and the U.S. Energy Research and Development Administration, Division of Geothermal Energy. The main emphasis of the program has been on locating sources for hot water convection systems and hot, dry, crystalline rock, to be utilized as space and process heating. The Arizona Bureau of Geology and Mineral Technology is compiling a special library on geothermal energy.

A preliminary map of the geothermal resources in Arizona was published in February 1978 by the Geological Survey Branch. It is a compilation of existing data printed on a U.S.G.S. 1:1,000,000 scale base map. It depicts hot springs (30°C; 86°F), cinder cones and extrusive volcanic rocks 3,000,000 years and younger, state and federally designated known geothermal resource areas, regions of high chemical geothermometers, high heat flow (τ 2.5 HFU), and moderate (τ 36°C/km; 108°F/mi) and high (τ 150°C/km; 181°F/mi) geothermal gradients.

Perhaps the most promising location, with bottom temperatures of 163°C (325°F) and 184°C (333°F) and discharge estimated at 19,000 1/min (5700 gal/min) at depths below 2 kilometers (1.2 miles), in two wells about 1 kilometer (0.6 mile) apart, is near Chandler in Maricopa County.

CHAPTER 16

ENVIRONMENTAL GEOLOGY

We are not all environmentalists—many are not even conservationists, however everyone becomes concerned whenever his immediate environment is threatened by some type of natural hazard. The most spectacular natural hazards we hear of are earthquakes, volcanic eruptions, and floods. In Arizona we are fortunate not to have severe earthquakes as in California, nor volcanic eruptions as in Washington, however there is potential for both. Floods are a frequent and predictable occurrence and occasional damage to man-made structures occurs as a result of landslides or rockfalls. Earth cracks are becoming more than a nuisance in some of the southern Arizona valleys due to excessive groundwater pumping (Figure 16–1). The study of such natural hazards is called environmental geology.

Figure 16-1. Earth crack in Mesa, formed by settling of unconsolidated valley fill as ground water is withdrawn.

Human populations typically ignore the warnings of geologists relative to hazards, prefering to believe in good old solid mother Earth. In reality, the intelligent use of the land requires an understanding of the potential natural hazards for any spot on the Earth's surface. Many hazards are relatively well known for various parts of Arizona. Historical records and earthquake monitoring equipment throughout the state records a "felt" earthquake frequency of approximately one per year, during the last 100 years (Figure 16–2). Certainly there has been very little property loss here, but there have been a few damaging quakes, especially in the Yuma area. The risk is high enough to require its recognition in the building codes of Phoenix, Tucson, Yuma, and other cities in the state.

Flooding of the usually dry major and minor drainages occurs seasonally in the population centers of metropolitan Phoenix, Tucson, and numerous small communities such as Duncan and Winslow, results in millions of dollars damage to roads, bridges, homes and farm lands, and occasional loss of life (Figure 16–2). The flood-prone areas are well known but, in spite of that knowledge, construction of buildings on these flood plains continues—and the flood damage scenario continues to be played out year after year.

The population centers continue to expand, in some cases, into terrain that carries the potential of damage from landslides and rockfalls, for example, the Camelback Mountain area in Phoenix. This risk is documented by such damage to many homes in the area.

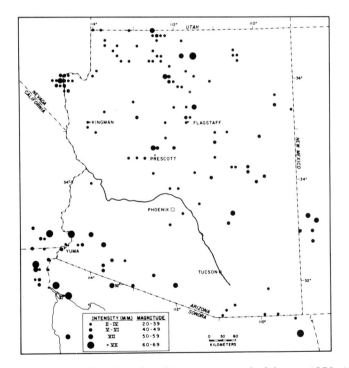

Figure 16–2. Location and intensity of earthquake occurrences in Arizona, 1852–1980. Modified from DuBois and Smith, 1980, Field notes of the Arizona Bureau of Geology and Mineral Technology, Volume 10, Number 3.

Land subsidence, although not entirely natural, is an increasing problem in many areas of heavy groundwater pumpage in the valleys around Phoenix and Tucson. An allied problem that is continually ignored is the rapid lowering of the groundwater table in those same areas, leading to an ultimate depletion of the water supply upon which those population centers depend.

Although irritating to many people, the evaluation of natural hazards must be incorporated in land-use planning, especially for large scale and environmentally sensitive construction projects such as dams, nuclear power plants, nuclear waste disposal and highways. Our planning has been typically short-sighted in the past, but the trend now seems to be toward increasing concern.

Figure 16–3. Flooding of Salt River and the resultant destruction of Mill Avenue between Tempe and Phoenix.

REFERENCES AND SUGGESTED READINGS

Anderson, R. Y. and J. W. Harshbarger (Editors), 1958. New Mexico Geological Society, Guidebook of the Black Mesa Basin, Ninth Field Conference, 205 p.

Anthony, J. W., S. A. Williams and R. A. Bideaux, 1977. Mineralogy of Arizona; University of Arizona Press, 255 p.

Arizona Bureau of Mines, 1969. Mineral and water resources of Arizona, Bulletin 180, University of Arizona Press, Tucson, Arizona, 638 p.

Arizona Geological Society Digest, 1976. Tectonic Digest, Volume Ten; Tucson, Arizona, 430 p.

Beus, S. S. and R. R. Rawson (Editors), 1979. Carboniferous stratigraphy in the Grand Canyon country, northern Arizona and southern Utah. Ninth International Congress of Carboniferous Stratigraphy and Geology, Field Trip No. 13, American Geological Institute Selected Guidebook Series No. 2, 138 p.

Billingsley, G. H., 1978. A synopsis of stratigraphy in the western Grand Canyon. Museum of Northern Arizona Research Paper 16, 27 p.

Breed W. J. and E. C. Roat (Editors), 1976. Geology of the Grand Canyon; Museum of Northern Arizona, 186 p.

Burt, D. M. and T. L. Pewe (Editors), 1978. Guidebook to the Geology of central Arizona; Arizona Bureau of Geology and Mineral Technology, Special Paper No. 2, 176 p.

Callender, J. F., J. C. Wilt and R. E. Clemons (Editors), 1978. Land of Cochise, southeastern Arizona; New Mexico Geological Society Guidebook, 372 p.

Conway, C. M., 1976. Petrology, structure, and evolution of a Precambrian volcanic and plutonic complex, Tonto Basin, Gila County, Arizona; unpublished Ph.D. Dissertation, California Institute of Technology, 460 p.

Dunning, C. H., 1959. Rocks to riches; Southwest Publishing Company, Phoenix, 406 p.

Fouch, T. D. and E. R. Magathan (Editors), 1980. Paleozoic paleogeography of the west-central United States, Rocky Mountain Paleogeography Symposium 1; Rocky Mountain Section of Society of Economic Paleontologists and Mineralogists, 431 p.

Four Corners Geological Society, 1969. Geology and natural history of the Grand Canyon region, Ninth Field Conference Guidebook, Powell Centennial River Expedition, 212 p.

Hale, M. E., 1976. Arizona energy, a framework for decision; University of Arizona Press, 160 p.

Huntoon, P. W., 1974, The post-Paleozoic structural geology of the eastern Grand Canyon, Arizona, *In* Breed, W. J. and E. C. Roat (Editors), Geology of the Grand Canyon.

James, H. L. (Editor), 1973. Guidebook of Monument Valley and vicinity; New Mexico Geological Society, 206. p.

Karlstrom, T. N. V., G. A. Swann and R. L. Eastwood, 1974. Geology of northern Arizona, with notes on archeology and paleoclimate; Part I, Regional Studies; Part II, Area Studies and Field Guides; Guidebook for Geological Society of America, Rocky Mountain Section Meeting, Flagstaff, Arizona, 805 p.

Keith, S. B., 1971. Geologic Guidebook 3, Highways of Arizona, Arizona Highways 85, 86 and 386; Arizona Bureau of Mines Bulletin 183, 80 p.

Kelley, V. C., 1958. Tectonics of the Black Mesa Basin region of Arizona, *In* Guidebook of the Black Mesa Basin, northeastern Arizona; New Mexico Geological Society, Ninth Field Conference.

Ludwig, K. R., 1974. Precambrian geology of the central Mazatzal Mountains, Arizona (Part I) and lead isotope heterogeneity in Precambrian igneous feldspars (Part II); unpublished Ph.D. Dissertation, California Institute of Technology, 363 p.

McKee, E. D. and C. E. Resser, 1945. Cambrian history of the Grand Canyon region; Carnegie Institution of Washington, Publication 563, 232 p.

McKee, E. D., R. F. Wilson, W. J. Breed and C. S. Breed, 1964. Evolution of the Colorado River in Arizona; Museum of Northern Arizona, 67 p.

Moore, R. T., 1972. Geology of the Virgin and Beaverdam Mountains; Arizona Bureau of Mines Bulletin 186, University of Arizona Press, Tucson, 65 p.

Newman, G. W. and H. D. Goode (Editors), 1979. Basin and Range Symposium, Rocky Mountain Association of Geologists and Utah Geological Association, 662 p.

Nininger, H. H., 1972. Find a falling star; Erikson Press, 254 p.

Peirce, H. W., 1967. Geologic Guidebook 2, Highways of Arizona, Arizona Highways 77 and 177; Arizona Bureau of Mines Bulletin 176, 73 p.

Peirce, H. W., S. B. Keith and J. C. Wilt, 1970. Coal, oil, natural gas, helium and uranium in Arizona; Arizona Bureau of Mines Bulletin 182, 289 p.

Pewe, T. L., 1974. Colorado River Guidebook: Lee's Ferry to Phantom Ranch, Tempe, Arizona, 79 p.

Pewe, T. L. and R. G. Updike, 1976. San Francisco Peaks: a guidebook to the geology; Museum of Northern Arizona, 80 p.

Rabbitt, M. C., E. D. McKee, C. B. Hunt and L. B. Leopold, 1969. The Colorado River region and John Wesley Powell; United States Geological Survey Professional Paper 669, 145 p.

Rigby, J. K., 1977. Southern Colorado Plateau Field Guide: Kendall/Hunt Publishing Company, 148 p.

Royse, C. F., M. F. Sheridan and H. W. Peirce, 1971. Geologic Guidebook 4, Highways of Arizona Highways 87, 88 and 188; Arizona Bureau of Mines Bulletin 184, 66 p.

Sheridan, M. F., 1978. Superstition wilderness guidebook: geology and trails, Tempe, Arizona, 54 p.

Shoemaker, E. M. and S. W. Kieffer, 1974. Guidebook to the geology of Meteor Crater, Arizona; Center for Meteorite Studies, Arizona State University, 66 p.

Shride, A. F., 1967. Younger Precambrian geology in southern Arizona. United States Geological Survey Professional Paper 566, 89 p.

Titley, S. R. (Editor), 1968. Southern Arizona Guidebook III; Arizona Geological Society, Tucson, 354 p.

Wilson, E. D., 1962. A Resume of the geology of Arizona; Arizona Bureau of Mines Bulletin 171, 140 p.

Wilson, E. D., 1965, Guidebook 1, Highways of Arizona, U.S. Highway 666; Arizona Bureau of Mines Bulletin 174, 68 p.

Wilt, J.C. and J. P. Jenney (Editors), 1976. Tectonic Digest; Arizona Geological Society Digest, Volume Ten, 420 p.

INDEX

Note—nonitalicized numbers refer to text pages; italicized numbers refer to figures.